Sinusoidal Three-Phase Windings
of Electric Machines

Jonas Juozas Buksnaitis

Sinusoidal Three-Phase Windings of Electric Machines

Jonas Juozas Buksnaitis
Institute of Energetics & Biotechnology
Aleksandras Stulginskis University
Kaunas, Lithuania

ISBN 978-3-319-82696-7   ISBN 978-3-319-42931-1   (eBook)
DOI 10.1007/978-3-319-42931-1

This Springer imprint is published by Springer Nature
The registered company is Springer International Publishing AG Switzerland

Preface

In five chapters of the monograph *Sinusoidal Three-Phase Windings of Electric Machines*, a comprehensive description of the following material is presented: (a) general theoretical foundations of sinusoidal three-phase windings, as well as creation of these windings with the maximum and average pitch with optimized pulsating and rotating magnetomotive force, and determination of number of turns in the coils of coil groups (Chap. 1); (b) determination of electromagnetic parameters of sinusoidal three-phase windings while changing the number of pole and phase slots (number of coils in the coil group) (Chap. 2); (c) calculation of magnetic circuit slot fill factor for all four types of previously created sinusoidal three-phase windings (Chap. 3); (d) creation of technological schemes for mechanized insertion of sinusoidal three-phase windings into the slots of magnetic circuit (Chap. 4); (e) determination and comparison of electromagnetic and energy-related parameters of factory-made motor with a single-layer preformed winding and rewound motor with a three-phase sinusoidal winding (Chap. 5).

In this monograph, the author performed a comprehensive analysis of four types of sinusoidal three-phase windings, as well as the theoretical investigation of related electromagnetic parameters; this investigation was also used as a basis to complete the qualitative evaluation of electromagnetic characteristics of discussed windings. How well did he perform this task is up to the monograph readers to decide.

The monograph is dedicated to a professional book, to the specialists in the field of electrical engineering, and could be used to deepen their knowledge and apply it in practice. Material can be also used as a source of scientific information in master's and doctoral studies.

The author is fully aware that he was unable to avoid all potential inaccuracies or other flaws in this edition. A part of these inconsistencies was eliminated while consulting Lithuanian specialists of electrical engineering. Additionally, the author wishes to express his gratitude to everyone who contributed to the manuscript preparation.

Kaunas, Lithuania   Jonas Juozas Buksnaitis

Introduction

In alternating current multiphase electrical machines, the serration of stator and rotor magnetic circuits and inconsistent distribution of windings, as well as other factors, create conditions for periodic non-sinusoidal rotating magnetic fields to form in the air gaps of these elements. Such instantaneous periodic functions of rotating magnetomotive force, distributed according to a non-sinusoidal law, can be expanded into rotating space harmonics of their direct or reverse sequence. Most often, the first (fundamental) space harmonic of rotating magnetomotive force performs useful dedicated functions in alternating current electrical machines. The impact of the higher-order space harmonics of rotating magnetomotive forces on the performance of such electrical machines is, essentially, negative: they increase power losses in electrical machines, deteriorate mechanical characteristics of induction motors, distort internal voltages which are induced in windings, create additional noises, resonance effects, etc. Each space harmonic of rotating magnetomotive force excites harmonics of internal voltage of the same order in stator and rotor windings, which in turn form non-sinusoidal internal voltage curves by adding up with each other.

In order to reduce or completely eliminate some of the higher-order space harmonics of rotating magnetomotive forces, i.e., to bring the space function of rotating magnetomotive force in the air gap of electrical machines, as well as the time function of voltage generated in windings, closer to sinusoidal distribution, certain measures are typically taken: coil span is reduced ($y<\tau$), windings are distributed ($q>1$), etc., where y—coil span; τ—pole pitch; q—number of stator slots (coils) per pole per phase. All these measures reduce harmonics of rotating magnetomotive forces and voltages induced by them. When the coil span y is reduced with respect to pole pitch τ of the fundamental harmonic, only some of the higher-order space harmonics of rotating magnetomotive forces are eliminated or reduced significantly. Space functions of rotating magnetomotive force in distributed windings have a characteristic staircase shape and are more similar to sinusoidal than square-shaped rotating magnetomotive force of concentrated winding.

Sometimes the coil turn numbers in distributed concentric single-phase winding coil groups, consisting of q coils and corresponding to a single winding pole, can be different when determined according to a certain law, i.e., $N_1 \neq N_2 \neq \ldots \neq N_i \neq \ldots \neq N_q$, where N_i—number of turns in i-th coil. The pulsating magnetomotive force space function, generated by such winding, is brought even closer to sinusoidal distribution. Therefore, the winding of this type is called a sinusoidal single-phase winding. Sinusoidal single-phase winding is a concentric alternating current winding consisting of uniform coil groups, the number of which in the phase winding matches the number of poles in this winding, while the numbers of turns in group-forming coils, which are distributed according to sinusoidal law starting from the symmetry axes of these groups, are different. The theory of single-phase sinusoidal windings is sufficiently substantiated, their calculations are well defined, and they have been used in single-phase induction motors for quite a long time. These motors with sinusoidal windings have noticeably better energy-related parameters and also include other good features.

However, there is not much material available in technical literature regarding sinusoidal three-phase windings; they are also not used to manufacture alternating current electrical machines. It can be asserted that the application of such three-phase windings, for example, in induction motors, could eliminate or reduce certain higher-order harmonics of rotating magnetomotive forces to minimum, thus improving their energy-related parameters. In this monograph a possibility to create several types of sinusoidal three-phase windings will be discussed. It is believed that windings of this type could substantially contribute to the improvement in alternating current electrical machines.

Contents

List of Main Symbols and Abbreviations

α_j	Width of the j-th rectangle of the stair-shaped rotating magnetomotive force curve half-period, expressed in electric degrees of the fundamental harmonic
β	Magnetic circuit slot pitch, expressed in electric degrees
F	Magnetomotive force
F_{m1}	Conditional amplitude value of the first (fundamental) harmonic of rotating magnetomotive force
$F_{m\nu}$	Conditional amplitude value of the ν-th harmonic of rotating magnetomotive force
F_{jr}	Conditional height of the j-th rectangle of stair-shaped rotating magnetomotive force curve half-period
f_ν	Absolute relative value of the amplitude of the ν-th harmonic of rotating magnetomotive force
i	Instantaneous electric current
k	Number of rectangles forming half-periods of the stair-shaped magnetomotive force curve
k_{w1}	Winding factor of the first harmonic
$k_{w\,\nu}$	Winding factor of the ν-th harmonic
k_{ef}	Winding electromagnetic efficiency factor
λ_{1pn}	Preliminary fill factor of the n-th magnetic circuit slot in maximum average pitch STW with optimized pulsating magnetomotive force
λ_{1rn}	Preliminary fill factor of the n-th magnetic circuit slot in maximum average pitch STW with optimized rotating magnetomotive force
λ_{2pn}	Preliminary fill factor of the n-th magnetic circuit slot in short average pitch STW with optimized pulsating magnetomotive force
λ_{2rn}	Preliminary fill factor of the n-th magnetic circuit slot in short average pitch STW with optimized rotating magnetomotive force
$\lambda_{1p\,n}^{*}$	Real fill factor of the n-th magnetic circuit slot in maximum average pitch STW with optimized pulsating magnetomotive force
$\lambda_{1r\,n}^{*}$	Real fill factor of the n-th magnetic circuit slot in maximum average pitch STW with optimized rotating magnetomotive force

$\lambda^*_{2\text{p }n}$ Real fill factor of the n-th magnetic circuit slot in short average pitch STW with optimized pulsating magnetomotive force

$\lambda^*_{2\text{r }n}$ Real fill factor of the n-th magnetic circuit slot in short average pitch STW with optimized rotating magnetomotive force

$\lambda^*_{1\text{p av}}$ Average fill factor of the magnetic circuit slots in maximum average pitch STW with optimized pulsating magnetomotive force

$\lambda^*_{1\text{r av}}$ Average fill factor of the magnetic circuit slots in maximum average pitch STW with optimized rotating magnetomotive force

$\lambda^*_{2\text{p av}}$ Average fill factor of the magnetic circuit slots in short average pitch STW with optimized pulsating magnetomotive force

$\lambda^*_{2\text{r av}}$ Average fill factor of the magnetic circuit slots in short average pitch STW with optimized rotating magnetomotive force

m Phase number

N_i Number of turns in the i-th coil

$N^*_{1\text{p }i}$ Relative number of turns in the i-th coil in maximum average pitch STW with optimized pulsating magnetomotive force

$N^*_{1\text{r }i}$ Relative number of turns in the i-th coil in maximum average pitch STW with optimized rotating magnetomotive force

$N^*_{2\text{p }i}$ Relative number of turns in the i-th coil in short average pitch STW with optimized pulsating magnetomotive force

$N^*_{2\text{r }i}$ Relative number of turns in the i-th coil in short average pitch STW with optimized rotating magnetomotive force

ν Number of space harmonic of magnetomotive force

p Number of pole pairs

q Number of stator slots (coils) per pole per phase

t Time

T Period

τ Pole pitch

$\upsilon_{1\text{p }i}$ Preliminary relative magnitude of i-th coil turn number in maximum average pitch STW with optimized pulsating magnetomotive force

$\upsilon_{1\text{r }i}$ Preliminary relative magnitude of i-th coil turn number in maximum average pitch STW with optimized rotating magnetomotive force

$\upsilon_{2\text{p }i}$ Preliminary relative magnitude of i-th coil turn number in short average pitch STW with optimized pulsating magnetomotive force

$\upsilon_{2\text{r }i}$ Preliminary relative magnitude of i-th coil turn number in short average pitch STW with optimized rotating magnetomotive force

y Coil span

Z Number of magnetic circuit slots

STW   Sinusoidal three-phase winding

O_{1q} Maximum average pitch double layer concentric (simple) three-phase winding

P_{1q} Maximum average pitch STW with optimized pulsating magnetomotive force

R_{1q} Maximum average pitch STW with optimized rotating magnetomotive force

O_{2q} Short average pitch double layer concentric (simple) three-phase winding

P_{2q} Short average pitch STW with optimized pulsating magnetomotive force

R_{2q} Short average pitch STW with optimized rotating magnetomotive force

Chapter 1
Fundamentals and Creation of Sinusoidal Three-Phase Windings (STW)

From the first look it could seem that to manufacture sinusoidal three-phase windings and to use them in practice should not cause any problems, as all types of three-phase windings with equal number of turns in their coils are made of single-phase windings with their starting points displaced in space by 120 electrical degrees. As it has been mentioned already, the theory of single-phase sinusoidal windings is well developed, calculations of these windings are established, and they have been used in single-phase induction motors for quite a long time. It may look that information related to these windings that has been accumulated over time should only be applied to distributed three-phase windings, and in this way to obtain single-layer or double-layer sinusoidal three-phase windings. However, it is not possible to accomplish that so easily due to multiple reasons. First of all, attention should be directed to the fact that not a single preformed single-layer or double-layer three-phase winding most commonly used in practice could be adapted for the creation of sinusoidal three-phase winding, since they do not fulfill fundamental structural conditions of electrical circuits of sinusoidal windings. These conditions are the following:

1. In each phase winding, positive and negative half-periods of pulsating magneto-motive forces with steps of different height could be induced in sinusoidal three-phase windings only by separate successively placed and respectively electromagnetically connected with each other groups of coils, i.e., these windings can only be formed of $6p$ equal coil groups; where p—number of pole pairs in sinusoidal three-phase winding;
2. In order to obtain space distribution of pulsating magnetomotive force close to sinusoidal with steps of different height at any time instant and symmetric in respect of coordinate axes, it is necessary that the coil groups determining this distribution and consisting of coils with different number of turns would be symmetric regarding the axes of considered coil groups in all aspects as well;
3. Spans of every coil group in sinusoidal three-phase windings have to be uniform and equal τ or $(\tau - 1)$; where τ—pole pitch expressed in slot pitch number.

© Springer International Publishing Switzerland 2016
J.J. Buksnaitis, *Sinusoidal Three-Phase Windings of Electric Machines*,
DOI 10.1007/978-3-319-42931-1_1

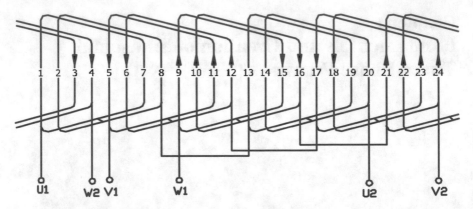

Fig. 1.1 Electrical diagram layout of four-pole single-layer preformed three-phase winding

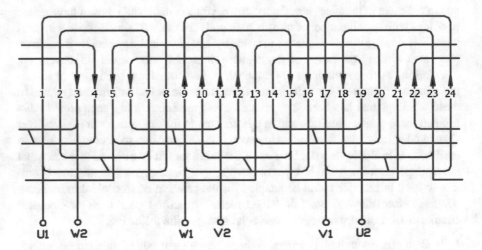

Fig. 1.2 Electrical diagram layout of four-pole concentric single-layer three-phase winding

Single-layer preformed three-phase windings do not match all three structural conditions of electrical circuits of sinusoidal windings (Fig. 1.1). These windings are formed of $3p$ coil groups only and each of them generates two-pole pulsating magnetic field. Furthermore, coil groups of these windings are not symmetric in respect of their axes and their span is larger than the pole pitch τ.

Single-layer concentric three-phase windings do not fulfill the first and the third structural conditions of electrical circuits of sinusoidal windings (Fig. 1.2). These windings are also formed of $3p$ coil groups only, and each of them generates two pole pulsating magnetic field. Additionally, span of coil groups in these windings is larger than the pole pitch τ as well.

Double-layer preformed three-phase windings do not fulfill the second structural condition of electrical circuits of sinusoidal windings (Fig. 1.3). Coil groups of these windings are not symmetric in respect of their axes.

The closest to sinusoidal three-phase winding is the concentric double-layer three-phase winding which matches all three structural conditions of electrical circuits of sinusoidal windings, as the coil group span in this type of winding can be equal to τ (maximum average pitch two-layer concentric three-phase winding) (Fig. 1.4) or $(\tau - 1)$ (short-pitch average pitch double-layer concentric three-phase winding) (Fig. 1.5).

Based on the theory of sinusoidal single-phase windings, we can conclude that sinusoidal three-phase windings can be created for distributed coil fed-in three-phase windings that are wound using flexible coils. In order not to violate three-phase winding symmetricity conditions, all phase coil groups of the considered winding have to be identical for both spatial placement of coils and dimensions of related coils and numbers of turns inside them. To achieve symmetricity of coil groups for each phase in respect of their axes, and to have the coil group span equal to τ or $(\tau - 1)$, only a single possible variant exists in case of considered winding, i.e., coils in each coil group have to be arranged concentrically. Sinusoidal three-phase windings with concentric coil groups have to be implemented as double-layer only, as just in such three-phase windings one group of coils excites a single-pole pulsating magnetic field. As it is known, in single-layer three-phase windings one group of coils excites two-pole pulsating magnetic field. Furthermore, significantly better winding distribution is achieved in double-layer three-phase windings, what will have even greater positive influence on sinusoidal three-phase windings. In fact, sinusoidal three-phase windings are a modification of double-layer concentric three-phase windings with identical number of coil turns. In other words, coil groups in the structure of electrical diagrams of the discussed sinusoidal three-phase windings, according to their creation nature, would not be much different from coil groups of sinusoidal single-phase windings.

Based on the provided structure of electrical diagrams of sinusoidal three-phase windings (Figs. 1.4 and 1.5), coils of each phase winding would be inserted into two-thirds of different magnetic circuit slots in all cases. In this respect, the maximum winding distribution can be achieved. Since sinusoidal windings can be created as double-layer only, it means that coils of a single phase winding, in fact, will take up only one-third of magnetic circuit slot number Z. This corresponds to one of existing fundamental requirements of three-phase windings. To have these windings symmetric in all aspects, half of active coil sides in each phase winding have to be placed into bottom layers of slots, and other side—into top layers.

Thr relation between the main parameters of sinusoidal three-phase windings is expressed using the same formula as for the non-sinusoidal three-phase windings:

$$Z = 2pmq \qquad\qquad (1.1)$$

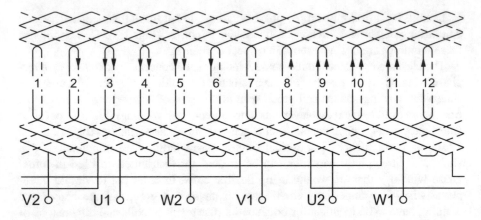

Fig. 1.3 Electrical diagram layout of preformed double-layer three-phase winding

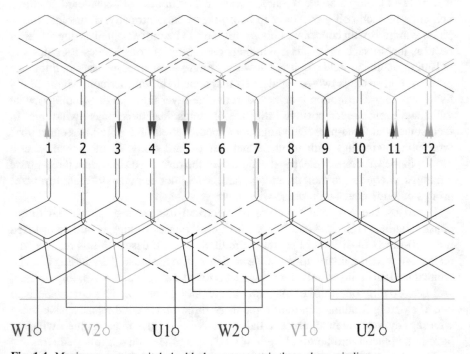

Fig. 1.4 Maximum average pitch double-layer concentric three-phase winding

where Z—number of slots in magnetic circuit of stator or rotor; $2p$—number of poles in sinusoidal three-phase winding; m—number of phase windings; q—number of stator slots (coils) per pole per phase.

It should be noted that the number of pole and phase coils (number of coils in the group of coils) in sinusoidal three-phase windings can only be integer. In this case, any two coil groups in the phase winding arranged side by side will excite symmetric pulsating magnetic fields of different polarity, which will be close to sinusoidal. Understandably, the greater the number of coils in coil groups q in these windings, the closer is the shape of instantaneous space function of pulsating magnetomotive force to sinusoidal. Therefore, the number of coils in coil groups q in sinusoidal three-phase windings has to be not less than two ($q=2; 3; 4; 5; ...$). When $q=1$, we would have a concentrated full-pitched three-phase winding only.

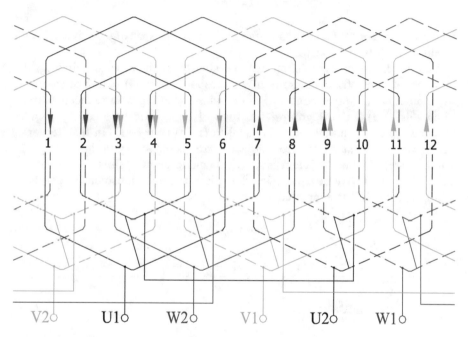

Fig. 1.5 Short-pitch average pitch double-layer concentric three-phase winding

The electrical diagram structure of the discussed maximum average pitch sinusoidal three-phase winding fully matches the electrical diagram structure of the double-layer concentric three-phase winding with its maximum pitch (Fig. 1.4). This sinusoidal winding is also created from coil groups, in which concentric coils have unequal span. The span of largest coils y_1, when the maximum average winding pitch is used, is equal to the pole pitch τ ($y_1 = \tau$). In this case, active sides of the largest coils from adjacent groups belonging to the same phase winding are laid in two layers into the same slots of magnetic circuit (slots 2; 4; 6; ...; 12) (Fig. 1.4). Spans of internal coils in coil groups, as in all types of concentric windings, is reduced by two slot pitches along the direction of central axes of these groups. Phase winding beginning and end terminal locations, coil and coil group connections in the considered three-phase winding also remain the same as in double-layer concentric three-phase winding with the maximum average winding pitch. For the discussed windings, this average pitch is expressed as:

$$y_{av_1} = \frac{y_1 + y_2 + \ldots + y_i + \ldots + y_q}{q}$$
$$= \frac{\tau + (\tau - 2) + \ldots + (\tau - 2(i-1)) + \ldots + (\tau - 2(q-1))}{q}$$
$$= \tau - q + 1 = 2\tau/3 + 1 = 2q + 1; \tag{1.2}$$

where y_1—span of the first coil in coil group ($y_1 = \tau$); y_i—span of the i-th coil in coil group; $\tau = Z/(2p)$—pole pitch.

For the discussed short-pitch average sinusoidal three-phase winding, the structure of its electrical diagram fully corresponds to the structure of electrical diagram of double-layer concentric three-phase winding with reduced average pitch (Fig. 1.5). In coil groups, the span of concentric side coils y_1, under reduced average winding pitch, is equal to $(\tau - 1)$ ($y_1 = \tau - 1$). In this scenario, active sides of the side coils from adjacent groups belonging to the same phase winding are laid into adjacent slots of magnetic circuit (slots 2–3; 4–5; …; 12–1) (Fig. 1.5). Average pitch in short-pitch average sinusoidal three-phase windings (Fig. 1.5) is expressed as:

$$y_{av_2} = \frac{y_1 + y_2 + \ldots + y_i + \ldots + y_q}{q}$$
$$= \frac{(\tau - 1) + (\tau - 3) + \ldots + (\tau - 2i + 1) + \ldots + (\tau - 2q + 1)}{q}$$
$$= \frac{2\tau}{3} = 2q. \tag{1.3}$$

It can be seen from formula (1.3) that the obtained average pitch of this winding is reduced by one-third of the pole pitch τ. This means that the higher-order space harmonics multiples of three will be equal to zero in pulsating magnetic fields generated by each phase winding. As it is known, these harmonics have zero magnitude in non-sinusoidal three-phase windings only in rotational magnetic fields, while in pulsating fields they mostly remain.

Other higher-order odd space harmonics of magnetic fields ($\nu = 5, 7, 11, \ldots$) will be significantly reduced or will be eliminated completely in sinusoidal three-phase windings after reducing winding span ($y_{av} < \tau$) and distributing these windings ($q \geq 2$), and also by winding coils in coil groups using different number of turns determined according to specific rules.

The beginnings of phase windings in sinusoidal three-phase windings will be arranged relative to each other in $2\pi/3$ electric radian steps, similarly as in conventional three-phase windings, i.e., displaced by $Z/(3p)$ stator slots. For example, assume that the beginning of phase winding U will be drawn out from the n-th slot of magnetic circuit. Then the beginning of phase winding V will have to be pulled out from $(Z/(3p)+n)$-th, and the beginning of phase winding W—from $(2Z/(3p)+n)$-th slot of magnetic circuit.

Based on the reasoning presented above, we have that it would not be possible to implement the discussed sinusoidal three-phase windings for entirely any numbers of magnetic circuit slots Z and pole pairs p. To obtain the pole and phase slot number $q \geq 2$, the number of magnetic circuit slots has to be even and a multiple of six. In Table 1.1, possible numbers of magnetic circuit slots are presented for which it is feasible (+) or not feasible (−) to construct sinusoidal three-phase windings with respective number of pole pairs matching the condition of $q \geq 2$.

It can be seen from Table 1.1 that when creating sinusoidal three-phase windings the numbers of magnetic circuit slots and pole pairs form a certain regularity. Sinusoidal three-phase windings with a single pole pair can be formed for each number of magnetic circuit slots; with two pole pairs—every second; with three—every third, etc., while starting to vary the number of stator slots (coils) per pole per phase beginning from two.

Table 1.1 Possible numbers of magnetic circuit slots and respective pole pairs matching condition $q \geq 2$ for creation of sinusoidal three-phase windings

Number of slots Z	Number of pole pairs p									
	1	2	3	4	5	6	7	8	9	...
12	+	−	−	−	−	−	−	−	−	
18	+	−	−	−	−	−	−	−	−	
24	+	+	−	−	−	−	−	−	−	
30	+	−	−	−	−	−	−	−	−	
36	+	+	+	−	−	−	−	−	−	
42	+	−	−	−	−	−	−	−	−	
48	+	+	−	+	−	−	−	−	−	
54	+	−	+	−	−	−	−	−	−	
60	+	+	−	−	+	−		−	−	
66	+	−	−	−	−	−	−	−	−	
72	+	+	+	+	−	+	−	−	−	
78	+	−	−	−	−	−	−	−	−	
84	+	+	−	−	−	−	+	−	−	
90	+	−	+	−	+	−	−	−	−	
96	+	+	−	+	−	−	−	+	−	
102	+	−	−	−	−	−	−	−	−	
108	+	+	+	−	−	+	−	−	+	
114	+	−	−	−	−	−	−	−	−	
...										

1.1   Creation of Maximum Average Pitch STW Through Optimization of Pulsating Magnetomotive Force

In sinusoidal three-phase windings with maximum average pitch, any phase winding connected to the alternating current power source should generate pulsating magnetic field, the space distribution of which would be more similar to sine function than space distributions of magnetomotive forces induced by phase windings of double-layer concentric three-phase winding. In this case, when analyzing the determination of the number of coil turns in maximum average pitch sinusoidal three-phase winding, it is sufficient to select from Fig. 1.6, a any single coil group of any phase winding, which determines the shape of space distribution of pulsating magnetomotive force.

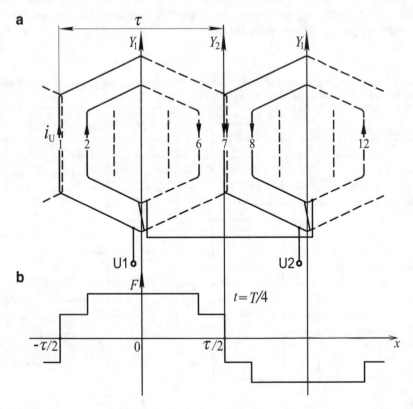

Fig. 1.6 Electrical diagram layout of a single phase (U) of the maximum average pitch concentric three-phase winding (**a**) and space distribution of its pulsating magnetomotive force in time $t = T/4$ (**b**)

Spatial position of instantaneous magnitudes of pulsating magnetomotive forces (Fig. 1.6b) does not change in time and their variation according to sinusoidal law will manifest itself in slots, e.g., 1; 2; 6; 7, in which the active coil sides of the same phase windings will be inserted. This means that the symmetry axes of these mag-netomotive forces in any scenario and in any moment of time will coincide with the corresponding coil group symmetry axes Y_1. These axes (Y_1) shall serve as reference axes when further examining the estimation of coil turn number. To the left or to the right side of the reference axes the span of coil groups corresponds to 90° electrical degrees or $Z/(4p) = \tau/2$ slot pitches.

Coils from any coil group are connected only in series and electric current of the same magnitude flows through them. Therefore, only the number of coils in their groups and number of turns in these coils influence the shape of space distributions of pulsating magnetomotive forces. It follows from here that in order to get the shape of space distribution of pulsating magnetomotive forces which would resem-ble the sine function the most, the number of coils turns in coil groups has to be distributed according to sinusoidal law in respect of the assumed reference axes Y_1.

Given the above considerations, the sine function values of the relevant angles are determined, which will be equal to the preliminary relative values of turn num-bers in respective coils:

$$
\begin{cases}
\upsilon_{1p\,1} = \sin(\pi/2) = 1 \; ; \\
\text{------------} \\
\upsilon_{1p\,i} = \sin\left[\pi/2 - \beta \cdot (i-1)\right] \; ; \\
\text{------------} \\
\upsilon_{1p\,q} = \sin\left[\pi/2 - \beta \cdot (q-1)\right] \; ;
\end{cases}
\tag{1.4}
$$

where $\beta = 2\pi p/Z$ is magnetic circuit slot pitch expressed in electrical radians; $i = 1 \div q$ is a coil number in a coil group.

The first number in a coil group is assigned to the coil with span $y_1 = \tau$, the second is assigned to the coil with span $y_2 = (\tau - 2)$, etc. Then, based on the equation system (1.4) we obtain that the first coil with the largest span will have the largest number of turns as well, while the q-th coil with the smallest span will have the minimum number of turns.

For the purpose of the further theoretical analysis, the sinusoidal three-phase winding of the maximum average pitch, after optimization of its pulsating magne-tomotive force, is associated with the concentrated full-pitched three-phase winding by converting the preliminary relative values of coil turn numbers obtained using expressions (1.4) to their real relative values:

$$
\begin{cases}
N_{1p\,1}^* = \upsilon_{1p1} / (2C_{1p}) = 1 / (2C_{1p}) \; ; \\
\text{------------} \\
N_{1p\,i}^* = \upsilon_{1p\,i} / (2C_{1p}); \\
\text{------------} \\
N_{1p\,q}^* = \upsilon_{1p\,q} / (2C_{1p});
\end{cases}
\tag{1.5}
$$

where $C_{1p} = \sum_{i=1}^{q} \upsilon_{1p\,i}$ is the sum of the preliminary relative values of coil turn numbers obtained from the equation system (1.4).

The sum of all the members from the equation system (1.5) has to match the following magnitude:

$$\sum_{i=1}^{q} N_{1pi}^{*} = 0.50. \tag{1.6}$$

This can be explained by the fact that the real relative value of the number of coil turns in the concentrated full-pitched three-phase winding is $N^{*}=1$, and the real relative value of number of turns in all coil group coils in double-layer distributed winding of any type $N^{**}=N^{*}/2=0.5$, because in these windings (non-sinusoidal) the relative value of single coil turn number is $N_{1}=N^{*}/(2q)=0.5/q$.

In Fig. 1.7, connection diagrams of coils and their groups for the maximum average pitch simple and sinusoidal concentric three-phase windings with $q=2$ are created.

In Fig. 1.4, electrical diagram of the analyzed three-phase windings is presented. The main parameters of these windings are: $2p=2$, $q=2$, $Z=12$, $\tau=6$, $y_{av}=5$, $\beta=30°$. Based on Eqs. (1.4), (1.5), i.e., through optimization of pulsating magnetomotive force, real relative values of coil turn number N_{1pi}^{*} of the discussed sinusoidal winding are calculated (Table 1.2).

In Fig. 1.8, connection diagrams of coils and their groups for the maximum average pitch simple and sinusoidal three-phase windings with $q=3$ are created.

Based on connection diagram (Fig. 1.8), electrical diagram of these three-phase windings is created, as shown in Fig. 1.9.

The main parameters of these windings are: $2p=2$, $q=3$, $Z=18$, $\tau=9$, $y_{av}=7$, $\beta=20°$. The real relative values of coil turn numbers N_{1pi}^{*} of this sinusoidal winding are calculated using Eqs. (1.4), (1.5) for optimization of the pulsating magnetomotive force (Table 1.3).

In Fig. 1.10, connection diagrams of coils and their groups for the maximum average pitch simple and sinusoidal three-phase windings with $q=4$ are created.

Based on connection diagram (Fig. 1.10), electrical diagram of these three-phase windings is created, as shown in Fig. 1.11.

The main parameters of these windings are: $2p=2$, $q=4$, $Z=24$, $\tau=12$, $y_{av}=9$, $\beta=15°$. Real relative values of coil turn numbers N_{1pi}^{*} of this sinusoidal winding are calculated using Eqs. (1.4), (1.5) for optimization of the pulsating magnetomotive force (Table 1.4).

In Fig. 1.12, connection diagrams of coils and their groups for the maximum average pitch simple and sinusoidal three-phase windings with $q=5$ are created.

Based on connection diagram (Fig. 1.12), electrical diagram of these three-phase windings is created, as shown in Fig. 1.13.

The main parameters of these windings are: $2p=2$, $q=5$, $Z=30$, $\tau=15$, $y_{av}=11$, $\beta=12°$. Real relative values of coil turn numbers N_{1pi}^{*} of this sinusoidal winding are calculated using Eqs. (1.4), (1.5) for optimization of the pulsating magnetomotive force (Table 1.5).

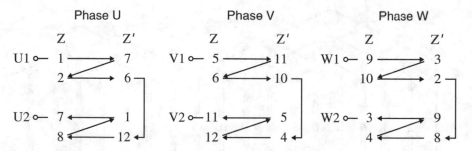

Fig. 1.7 Connection diagrams of coils and their groups for the maximum average pitch simple and STW

Table 1.2 Real relative values of coil turn number in coil group for maximum average pitch simple and STW with optimized pulsating magnetomotive force with $q=2$

	Winding type	
Number of coil in coil group	Simple (O_{12})	Sinusoidal (P_{12})
1	0.250	**0.268**
2	0.250	**0.232**

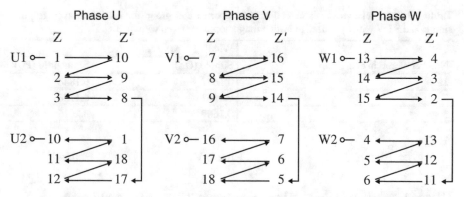

Fig. 1.8 Connection diagrams of coils and their groups for the maximum average pitch simple and STW

In Fig. 1.14, connection diagrams of coils and their groups for the maximum average pitch simple and sinusoidal three-phase windings with $q=6$ are created.

Based on connection diagram (Fig. 1.14), electrical diagram of these three-phase windings is created, as shown in Fig. 1.15.

The main parameters of these windings are: $2p=2$, $q=6$, $Z=36$, $\tau=18$, $y_{av}=13$, $\beta=10°$. Real relative values of coil turn numbers N_{1pi}^{*} of this sinusoidal winding are calculated using Eqs. (1.4), (1.5) for optimization of the pulsating magnetomotive force (Table 1.6).

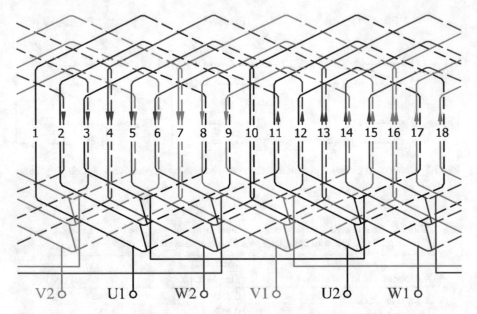

Fig. 1.9 Electrical diagram layout of maximum average pitch simple and STW with $q=3$

Table 1.3 Real relative values of coil turn number in coil group for maximum average pitch simple and STW with optimized pulsating magnetomotive force with $q=3$

Number of coil in coil group	Winding type	
	Simple (O_{13})	**Sinusoidal (P_{13})**
1	0.1667	**0.1848**
2	0.1667	**0.1736**
3	0.1667	**0.1416**

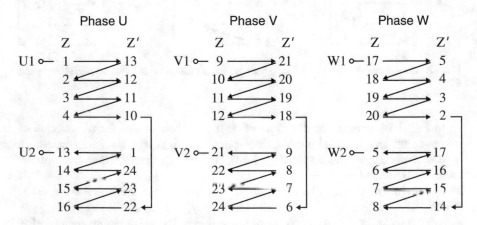

Fig. 1.10 Connection diagrams of coils and their groups for the maximum average pitch simple and STW

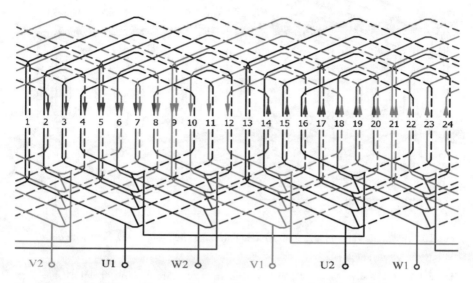

Fig. 1.11 Electrical diagram layout of maximum average pitch simple and STW with $q=4$

Table 1.4 Real relative values of coil turn number in coil group for maximum average pitch simple and STW with optimized pulsating magnetomotive force with $q=4$

Number of coil in coil group	Winding type	
	Simple (O_{14})	Sinusoidal (P_{14})
1	0.1250	**0.1413**
2	0.1250	**0.1365**
3	0.1250	**0.1223**
4	0.1250	**0.0999**

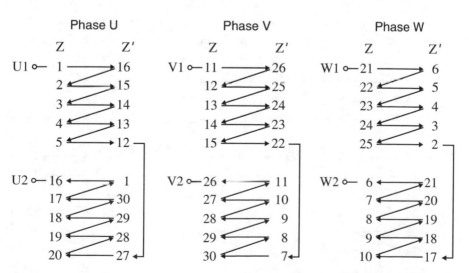

Fig. 1.12 Connection diagrams of coils and their groups for the maximum average pitch simple and STW

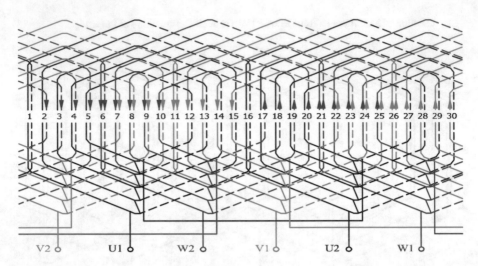

Fig. 1.13 Electrical diagram layout of maximum average pitch simple and STW with $q=5$

Table 1.5 Real relative values of coil turn number in coil group for maximum average pitch simple and STW with optimized pulsating magnetomotive force with $q=5$

Number of coil in coil group	Winding type	
	Simple (O_{15})	Sinusoidal (P_{15})
1	0.10	**0.1144**
2	0.10	**0.1119**
3	0.10	**0.1045**
4	0.10	**0.0926**
5	0.10	**0.0766**

Phase U

Z   Z'

U1 ○— 1 → 19
2 → 18
3 → 17
4 → 16
5 → 15
6 → 14

U2 ○— 19 → 1
20 → 36
21 → 35
22 → 34
23 → 33
24 — 32

Phase V

Z   Z'

V1 ○— 13 → 31
14 → 30
15 → 29
16 → 28
17 → 27
18 → 26

V2 ○— 31 → 13
32 → 12
33 → 11
34 → 10
35 → 9
36 — 8

Phase W

Z   Z'

W1 ○— 25 → 7
26 → 6
27 → 5
28 → 4
29 → 3
30 → 2

W2 ○— 7 → 25
8 → 24
9 → 23
10 → 22
11 → 21
12 — 20

Fig. 1.14 Connection diagrams of coils and their groups for the maximum average pitch simple and STW

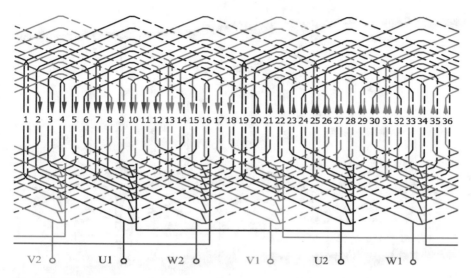

Fig. 1.15 Electrical diagram layout of maximum average pitch simple and STW with $q=6$

Table 1.6 Real relative values of coil turn number in coil group for maximum average pitch simple and STW with optimized pulsating magnetomotive force with $q=6$

Number of coil in coil group	Winding type	
	Simple (O_{16})	Sinusoidal (P_{16})
1	0.0833	0.0962
2	0.0833	0.0947
3	0.0833	0.0904
4	0.0833	0.0833
5	0.0833	0.0737
6	0.0833	0.0618

1.2   Creation of Maximum Average Pitch STW Through Optimization of Rotating Magnetomotive Force

By summing the optimized instantaneous space functions of pulsating magneto-motive forces of all three-phase windings in fixed moments of time, when the axes of these functions are shifted in space every 120° electrical degrees, the space functions of rotating magnetomotive forces in the same moments of time are obtained (Fig. 1.16b). But such functions calculated in this way are not the closest to the ideal sinusoid. It can be explained by the fact that the space distribution of the pulsating magnetomotive force if shaped by the group of coils consisting of q coils. As space functions of pulsating and rotating magnetomotive forces both in simple and sinusoidal double-layer concentric three-phase windings are

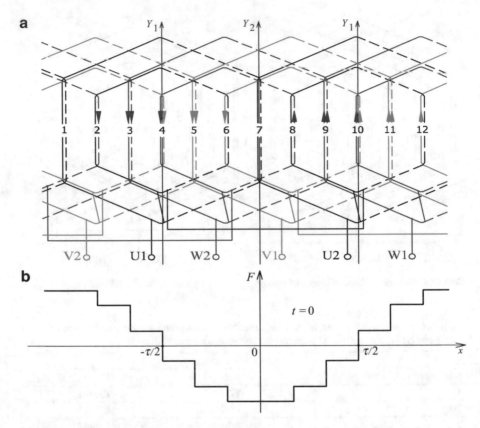

Fig. 1.16 Electrical diagram layout of the maximum average pitch concentric three-phase winding with $q=2$ (**a**) and space distribution of rotating magnetomotive force of this winding in time $t=0$ (**b**)

symmetric in respect of the coordinate axes, it is enough to consider only a quarter of the period T of these functions when analyzing related magnetomotive force. Accordingly, the pulsating magnetomotive force space distribution of width $\tau/2$ is shaped by the active sides of q coils. This distribution of rotating magnetomotive force is formed in the analyzed windings by the active coil sides located in the $Z/4p$ slots of the stator magnetic circuit. In three-phase windings, the magnitude $Z/4p$ $((2pmq)/4p=3q/2)$ is 1.5 times larger than q. The larger the number of the winding-containing magnetic circuit slots is involved in the formation of the 1/4 period distribution of pulsating or rotating magnetomotive force in the air gap, the more its function resembles sinusoidal. Therefore, during the optimization of the rotating magnetomotive force space distribution, it can be brought closer to sinusoidal function.

The directions of electric currents in the winding layout diagram in Fig. 1.16a, are presented on the basis of the phasor diagram of the three-phase current in time $t=0$ (Fig. 1.17).

Fig. 1.17 Phasor diagram
of the three-phase current
in time $t=0$

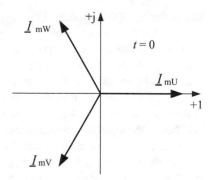

Analytical expressions of currents (Fig. 1.17) in time $t=0$:

$$\begin{cases} i_U = I_{mU} \sin\omega t = I_{mU} \sin 0^\circ = 0; \\ i_V = I_{mV} \sin\left(\omega t - 120^\circ\right) = I_{mV} \sin\left(-120^\circ\right) = -0.866 I_{mV}; \\ i_W = I_{mW} \sin\left(\omega t - 240^\circ\right) = I_{mW} \sin\left(-240^\circ\right) = 0.866 I_{mW}. \end{cases} \quad (1.7)$$

Based on Fig. 1.16, the reference axes are defined. They coincide with the symmetry axes of half-periods of rotating magnetomotive force (Fig. 1.16b), i.e., pass through the symmetry axes of the first or seventh slot of magnetic circuit (Fig. 1.16, axis Y_2). In these slots, the active sides of coils of the same phase (U) and largest span y_1 are located. Electric current does not flow through these coils in time $t=0$ ($i_U = 0$). In the same instant of time the current also does not flow through other coils of phase U located in magnetic circuit slots to the left or to the right side of the selected reference axes. In fixed moment of time the flow of currents is present only in coils of other phases (V and W). This fact greatly facilitates the optimization of rotating magnetomotive force when determining the preliminary relative values of coil turn numbers in coil groups.

In order to obtain the distribution of rotating magnetomotive force in time $t = 0$ as close to sinusoidal as possible, the relative values of number of turns in coils from coil groups must be determined according to the sine function of the space coordinate with the origin that can match any previously indicated reference axis (Y_2). Given the aforesaid considerations, the sine function values of the relevant angles expressed in electrical degrees and which will be equal to the preliminary relative values of number of coil turns are determined:

$$\begin{cases} \upsilon_{1r1} = \sin\left(\beta \cdot q\right)/2; \\ \upsilon_{1r2} = \sin\left[\beta\left(q-1\right)\right] \\ \text{-----------} \\ \upsilon_{1ri} = \sin\left[\beta\left(q+1-i\right)\right]; \\ \text{-----------} \\ \upsilon_{1rq} = \sin\beta; \end{cases} \quad (1.8)$$

where $\beta = 2\pi p/Z$ is magnetic circuit slot pitch expressed in electrical radians; $i = 1 \div q$ is a coil number in a coil group.

The first number in a coil group is assigned to the coil with span $y_1 = \tau$, the second is assigned to the coil with span $y_2 = (\tau - 2)$, etc. Then, based on the equation system (1.8) we obtain that in coil groups the first coil with the largest span will not have the largest number of turns, as coils of such span which belong to the same phase windings fill the same slots (1, 3, 5; 7; etc.) (Fig. 1.16). Second coil in coil groups will have the largest number of turns, while the q-th coil with the smallest span will have the minimum number of turns.

For the purpose of the further theoretical analysis, the sinusoidal three-phase winding of the maximum average pitch, after optimization of its rotating magnetomotive force, is associated with the concentrated full-pitched three-phase winding by converting the preliminary relative values of coil turn numbers obtained using expressions (1.7) to their real relative values:

$$
\begin{cases}
N_{1r1}^* = \upsilon_{1r1} / (2C_{1r}); \\
N_{1r2}^* = \upsilon_{1r2} / (2C_{1r}) \\
----------\\
N_{1ri}^* = \upsilon_{1ri} / (2\ C_{1r}); \\
----------\\
N_{1rq}^* = \upsilon_{1rq} / (2\ C_{1r});
\end{cases} \tag{1.9}
$$

where $C_{1r} = \sum\limits_{i=1}^{q} \upsilon_{1r\,i}$ is the sum of the preliminary relative values of coil turn numbers obtained from the equation system (1.8).

The sum of all the members N_{1ri}^* calculated from the equation system (1.9) has to match the magnitude of 0.50, the same like when optimizing pulsating magnetomotive force.

Based on parameters of created windings with $q=2$ (shown in Figs. 1.4 and 1.7) and using Eqs. (1.8) and (1.9) for optimization of the rotating magnetomotive force, the real relative values of coil turn numbers N_{1ri}^* of the discussed sinusoidal winding are calculated (Table 1.7).

Based on parameters of created windings with $q=3$ (shown in Figs. 1.8 and 1.9) and using Eqs. (1.8) and (1.9) for optimization of the rotating magnetomotive force, the real relative values of coil turn numbers N_{1ri}^* of the discussed sinusoidal winding are calculated (Table 1.8).

Based on parameters of created windings with $q=4$ (shown in Figs. 1.10 and 1.11) and using Eqs. (1.8) and (1.9) for optimization of the rotating magnetomotive force, the real relative values of coil turn numbers N_{1ri}^* of the discussed sinusoidal winding are calculated (Table 1.9).

Based on parameters of created windings with $q=5$ (shown in Figs. 1.12 and 1.13) and using Eqs. (1.8) and (1.9) for optimization of the rotating magnetomotive force, the real relative values of coil turn numbers N_{1ri}^* of the discussed sinusoidal winding are calculated (Table 1.10).

Table 1.7 Real relative values of coil turn number in coil group for maximum average pitch simple and STW with optimized rotating magnetomotive force with $q = 2$

Number of coil in coil group	Winding type	
	Simple (O_{12})	Sinusoidal (R_{12})
1	0.250	**0.232**
2	0.250	**0.268**

Table 1.8 Real relative values of coil turn number in coil group for maximum average pitch simple and STW with optimized rotating magnetomotive force with $q = 3$

Number of coil in coil group	Winding type	
	Simple (O_{13})	Sinusoidal (R_{13})
1	0.1667	**0.1527**
2	0.1667	**0.2267**
3	0.1667	**0.1206**

Table 1.9 Real relative values of coil turn number in coil group for maximum average pitch simple and STW with optimized rotating magnetomotive force with $q = 4$

Number of coil in coil group	Winding type	
	Simple (O_{14})	Sinusoidal (R_{14})
1	0.1250	**0.1140**
2	0.1250	**0.1862**
3	0.1250	**0.1317**
4	0.1250	**0.0681**

Table 1.10 Real relative values of coil turn number in coil group for maximum average pitch simple and STW with optimized rotating magnetomotive force with $q = 5$

Number of coil in coil group	Winding type	
	Simple (O_{15})	Sinusoidal (R_{15})
1	0.10	**0.0910**
2	0.10	**0.1562**
3	0.10	**0.1236**
4	0.10	**0.0855**
5	0.10	**0.0437**

Based on parameters of created windings with $q = 6$ (shown in Figs. 1.14 and 1.15) and using Eqs. (1.8) and (1.9) for optimization of the rotating magnetomotive force, the real relative values of coil turn numbers N_{1ri}^* of the discussed sinusoidal winding are calculated (Table 1.11).

Table 1.11 Real relative values of coil turn number in coil group for maximum average pitch simple and STW with optimized rotating magnetomotive force with $q=6$

Number of coil in coil group	Winding type	
	Simple (O_{16})	Sinusoidal (R_{16})
1	0.0833	0.0758
2	0.0833	0.1340
3	0.0833	0.1125
4	0.0833	0.0875
5	0.0833	0.0598
6	0.0833	0.0304

1.3 Creation of Short Average Pitch STW Through Optimization of Pulsating Magnetomotive Force

In sinusoidal three-phase winding with reduced average pitch, any phase winding connected to the source of alternating current should generate a pulsating magnetic field, and the space distribution of this field would be more similar to sine function than space distributions of pulsating magnetomotive forces induced by phase windings of simple concentric three-phase winding of the same pitch. In this case, when analyzing the determination of the number of coil turns in short average pitch sinusoidal three-phase winding, it is sufficient to select from Fig. 1.18, a any single coil group of any phase winding, which determines the shape of space distribution of pulsating magnetomotive force.

As coils in coil groups are connected only in series, the alternating electric current of the same magnitude flows through them, and therefore it does not affect the shape of pulsating magnetic field distribution in the air gap. Then, to have the shape of half-period of pulsating magnetomotive force generated by each group of coils as close to sinusoidal as possible, the number of coil turns in these groups of coils has to vary according to sinusoidal law in respect of the reference axes Y_1.

Given the above considerations, the sine function values of the relevant angles expressed in electrical radians are determined, which will be equal to the preliminary relative values of turn numbers in respective coils:

$$\begin{cases} \upsilon_{2p\,1} = \sin\left[(\pi-\beta)/2\right] ; \\ \text{-------------} \\ \upsilon_{2p\,i} = \sin\left[(\pi-\beta)/2 - \beta\cdot(i-1)\right] ; \\ \text{-------------} \\ \upsilon_{2p\,q} = \sin\left[(\pi-\beta)/2 - \beta\cdot(q-1)\right] ; \end{cases} \qquad (1.10)$$

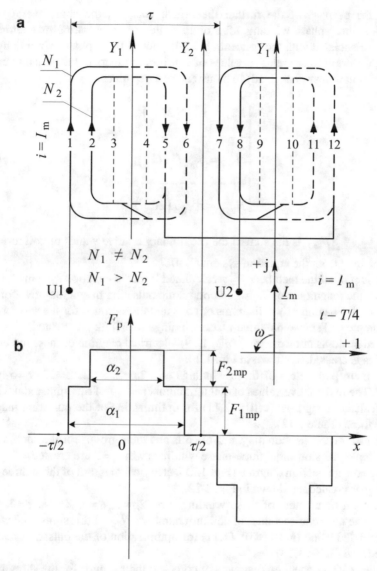

Fig. 1.18 Electrical diagram layout of a single phase (U) of the short average pitch concentric three-phase winding (**a**) and space distribution of its pulsating magnetomotive force in time $t=T/4$ (**b**)

where $\beta=2\pi p/Z$ is magnetic circuit slot pitch expressed in electrical radians; $i=1\div q$ is a coil number in a coil group.

The first number in a coil group is assigned to the coil with span $y_1=(\tau-1)$, the second is assigned to the coil with span $y_2=(\tau-3)$, etc. Then, based on the equation system (1.10) we obtain that in coil groups the first coil with the largest span will have the highest number of turns as well, while q-th coil of the smallest span will have the lowest number of turns.

For the purpose of the further theoretical analysis, the short average pitch sinusoidal three-phase winding, after optimization of its pulsating magnetomotive force, is associated with the concentrated full-pitched three-phase winding by converting the preliminary relative values of coil turn numbers obtained using expressions of equation system (1.10) to their real relative values:

$$
\begin{cases}
N^*_{2p\,1} = \upsilon_{2p1} / \left(2C_{2p} \right) ; \\
\text{-----------} \\
N^*_{2p\,i} = \upsilon_{2p\,i} / \left(2C_{2p} \right); \\
\text{-----------} \\
N^*_{2p\,q} = \upsilon_{2p\,q} / \left(2C_{2p} \right);
\end{cases}
\tag{1.11}
$$

where $C_{2p} = \sum\limits_{i=1}^{q} \upsilon_{2p\,i}$ is the sum of the preliminary relative values of coil turn numbers obtained from the equation system (1.10).

The sum of all the members N^*_{2pi} calculated from the equation system (1.11) has to match the magnitude of 0.5 when optimizing pulsating magnetomotive force.

In Fig. 1.19, connection diagrams of coils and their groups for the short average pitch simple and sinusoidal three-phase windings with $q=2$ are created.

Based on connection diagram (Fig. 1.19), electrical diagram of these three-phase windings is created, as shown in Fig. 1.20.

The main parameters of these windings are: $2p=2$, $q=2$, $Z=12$, $\tau=6$, $y_{av}=4$, $\beta=30°$. The real relative values of coil turn numbers N^*_{2pi} of this sinusoidal winding are calculated using Eqs. (1.10), (1.11) for optimization of the pulsating magnetomotive force (Table 1.12).

In Fig. 1.21, connection diagrams of coils and their groups for the short average pitch simple and sinusoidal three-phase windings with $q=3$ are created.

Based on connection diagram (Fig. 1.21), electrical diagram of these three-phase windings is created, as shown in Fig. 1.22.

The main parameters of these windings are: $2p=2$, $q=3$, $Z=18$, $\tau=9$, $y_{av}=6$, $\beta=20°$. The real relative values of coil turn numbers N^*_{2pi} of this sinusoidal winding are calculated using Eqs. (1.10), (1.11) for optimization of the pulsating magnetomotive force (Table 1.13).

In Fig. 1.23, connection diagrams of coils and their groups for the short average pitch simple and sinusoidal three-phase windings with $q=4$ are created.

Based on connection diagram (Fig. 1.23), electrical diagram of these three-phase windings is created, as shown in Fig. 1.24.

The main parameters of these windings are: $2p=2$, $q=4$, $Z=24$, $\tau=12$, $y_{av}=8$, $\beta=15°$. The real relative values of coil turn numbers N^*_{2pi} of this sinusoidal winding are calculated using Eqs. (1.10), (1.11) for optimization of the pulsating magnetomotive force (Table 1.14).

In Fig. 1.25, connection diagrams of coils and their groups for the short average pitch simple and sinusoidal three-phase windings with $q=5$ are created.

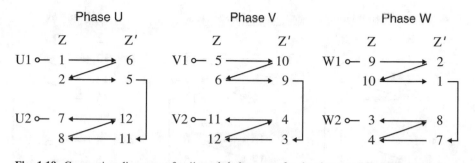

Fig. 1.19 Connection diagrams of coils and their groups for the short average pitch simple and STW

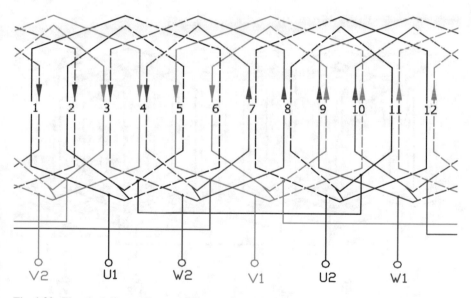

Fig. 1.20 Electrical diagram layout of short average pitch simple and STW with $q=2$

Table 1.12 Real relative values of coil turn number in coil group for short average pitch simple and STW with optimized pulsating magnetomotive force with $q=2$

Number of coil in coil group	Winding type	
	Simple (O_{22})	**Sinusoidal (P_{22})**
1	0.250	**0.264**
2	0.250	**0.236**

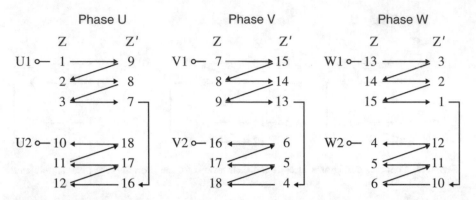

Fig. 1.21 Connection diagrams of coils and their groups for the short average pitch simple and STW

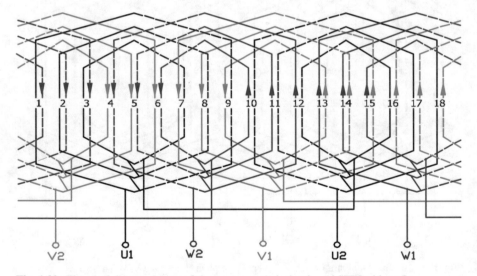

Fig. 1.22 Electrical diagram layout of short average pitch simple and STW with $q=3$

Table 1.13 Real relative values of coil turn number in coil group for short average pitch simple and STW with optimized pulsating magnetomotive force with $q=3$

Number of coil in coil group	Winding type	
	Simple (O_{23})	**Sinusoidal (P_{23})**
1	0.1667	**0.1975**
2	0.1667	**0.1736**
3	0.1667	**0.1289**

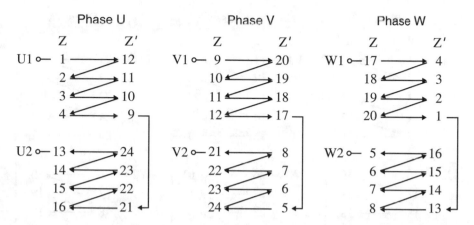

Fig. 1.23 Connection diagrams of coils and their groups for the short average pitch simple and STW

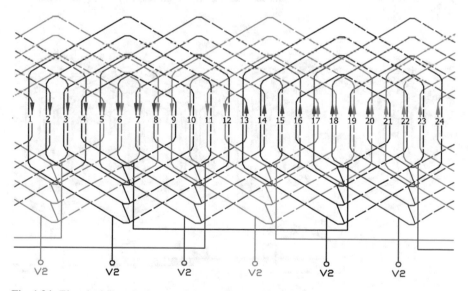

Fig. 1.24 Electrical diagram layout of short average pitch simple and STW with $q=4$

Table 1.14 Real relative values of coil turn number in coil group for short average pitch simple and STW with optimized pulsating magnetomotive force with $q=4$

Number of coil in coil group	Winding type	
	Simple (O_{24})	Sinusoidal (P_{24})
1	0.1250	0.1494
2	0.1250	0.1392
3	0.1250	0.1196
4	0.1250	0.0918

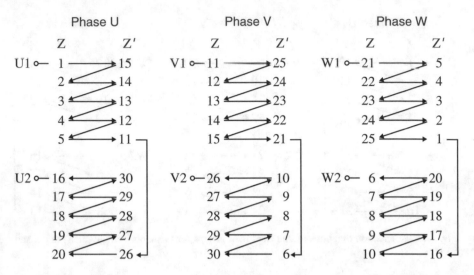

Fig. 1.25 Connection diagrams of coils and their groups for the short average pitch simple and STW

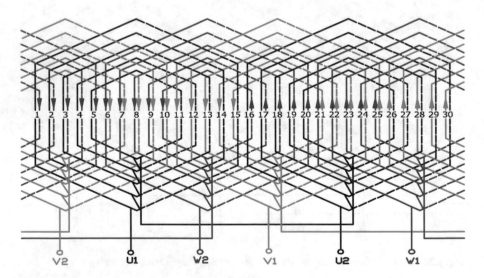

Fig. 1.26 Electrical diagram layout of short average pitch simple and STW with $q=5$

Based on connection diagram (Fig. 1.25), electrical diagram of these three-phase windings is created, as shown in Fig. 1.26.

The main parameters of these windings are: $2p=2$, $q=5$, $Z=30$, $\iota-15$, $y_{av}=10$, $\beta=12°$. The real relative values of coil turn numbers N^*_{2pi} of this sinusoidal winding are calculated using Eqs. (1.10), (1.11) for optimization of the pulsating magneto-motive force (Table 1.15).

Table 1.15 Real relative values of coil turn number in coil group for short average pitch simple and STW with optimized pulsating magnetomotive force with $q=5$

Number of coil in coil group	Winding type	
	Simple (O_{25})	Sinusoidal (P_{25})
1	0.10	**0.1200**
2	0.10	**0.1148**
3	0.10	**0.1045**
4	0.10	**0.0897**
5	0.10	**0.0710**

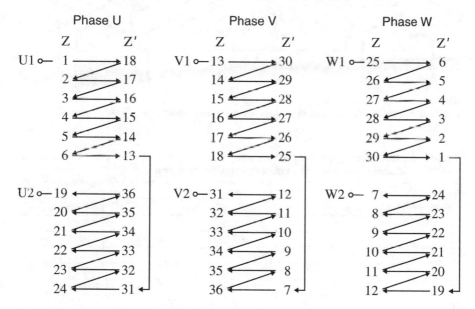

Fig. 1.27 Connection diagrams of coils and their groups for the short average pitch simple and STW

In Fig. 1.27, connection diagrams of coils and their groups for the short average pitch simple and sinusoidal three-phase windings with $q=6$ are created.

Based on connection diagram (Fig. 1.27), electrical diagram of these three-phase windings is created, as shown in Fig. 1.28.

The main parameters of these windings are: $2p=2$, $q=6$, $Z=36$, $\tau=18$, $y_{av}=12$, $\beta=10°$. The real relative values of coil turn numbers N_{2pi}^* of this sinusoidal winding are calculated using Eqs. (1.10), (1.11) for optimization of the pulsating magnetomotive force (Table 1.16).

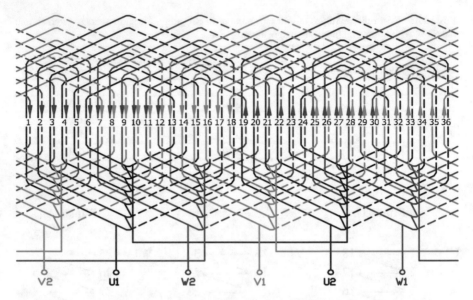

Fig. 1.28 Electrical diagram layout of short average pitch simple and STW with $q=6$

Table 1.16 Real relative values of coil turn number in coil group for short average pitch simple and STW with optimized pulsating magnetomotive force with $q=6$

	Winding type	
Number of coil in coil group	Simple (O_{26})	**Sinusoidal (P_{26})**
1	0.0833	**0.1003**
2	0.0833	**0.0972**
3	0.0833	**0.0912**
4	0.0833	**0.0824**
5	0.0833	**0.0712**
6	0.0833	**0.0577**

1.4   Creation of Short Average Pitch STW Through Optimization of Rotating Magnetomotive Force

The space distributions of rotating magnetomotive forces obtained from short average pitch sinusoidal three-phase windings after optimization of their pulsating magnetomotive forces are not optimal. In order to bring the shape of rotating magnetomotive forces closer to sinusoidal distribution even further, it is necessary to discuss the concentric three-phase windings which create these magnetomotive forces in their entirety.

From Fig. 1.29 we can see that the symmetry axes of half-periods of rotating magnetomotive forces obtained in time $t=0$ coincide with the symmetry axes Y_2 of two adjacent coil groups of phase U.

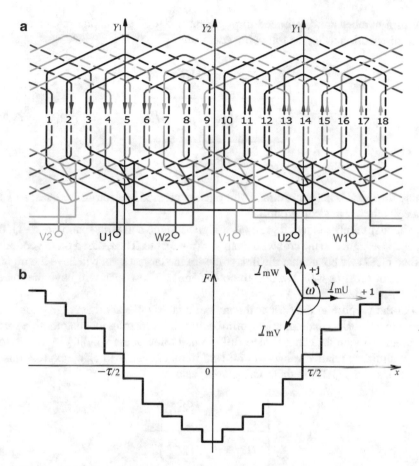

Fig. 1.29 Electrical diagram layout of a simple concentric three-phase winding (**a**) and space distribution of rotating magnetomotive force generated by this winding in time $t = 0$ (**b**)

These axes are assumed as the initial reference axes when optimizing the rotating magnetomotive force. It is necessary to draw attention to the fact that in q slots to the right or to the right side of the axis Y_2 in time instant $t = 0$ the electric current flows only through single layers of the winding (slots 7, 8, 9, 10, 11, and 12) belonging to phases V and W, while in other layers of the same slots belonging to phase U its magnitude is zero. Thus, the peak of the rotating magnetomotive force wave over $2q$ slots is shaped by V and W phase coil groups only (by one phase to one side from the axis Y_2, and by another phase to another side). The coils in coil groups of phase V and W windings are arranged symmetrically in respect of the Y_2 axis. Then, in order to create the shape of rotating magnetomotive force in time $t = 0$ as close to sinusoidal distribution as possible, the number of coil turns in these coil groups has to be determined according to the sine function of the space coordinate originating from Y_2 axis. Given the aforesaid considerations, the preliminary relative values of coil turn number are determined through optimization of rotating magnetomotive force distribution using the following equations:

$$
\begin{cases}
\upsilon_{2r\,1} = \sin\left[(q-1)\beta + \beta/2\right] ; \\
\text{------------} \\
\upsilon_{2r\,i} = \sin\left[(q-i)\beta + \beta/2\right] ; \\
\text{------------} \\
\upsilon_{2r\,q} = \sin(\beta/2) ;
\end{cases}
\tag{1.12}
$$

where $\beta = 2\pi p/Z$ is magnetic circuit slot pitch expressed in electrical radians; $i = 1 \div q$ is a coil number in a coil group.

The first number in a coil group is assigned to the coil with span $y_1 = (\tau - 1)$, the second is assigned to the coil with span $y_2 = (\tau - 3)$, etc. Then, based on the equation system (1.12) we obtain that the first coil with the largest span will have the highest number of turns as well, while q-th coil of the smallest span will have the lowest number of turns.

For the purpose of the further theoretical analysis, the short average pitch sinusoidal three-phase winding, after optimization of its rotating magnetomotive force, is associated with the concentrated full-pitched three-phase winding by converting the preliminary relative values of coil turn numbers obtained using expressions of equation system (1.12) to their real relative values:

$$
\begin{cases}
N^*_{2r1} = \upsilon_{2r1} / (2C_{2r}); \\
\text{------------} \\
N^*_{2ri} = \upsilon_{2ri} / (2C_{2r}); \\
\text{------------} \\
N^*_{2rq} = \upsilon_{2rq} / (2C_{2r});
\end{cases}
\tag{1.13}
$$

where $C_{2r} = \sum\limits_{i=1}^{q} \upsilon_{2r\,i}$ is the sum of the preliminary relative values of coil turn numbers obtained from the equation system (1.12).

The sum of all the members N^*_{2ri} calculated from the equation system (1.13) has to match the magnitude of 0.50 when optimizing rotating magnetomotive force.

Based on parameters of created windings with $q=2$ (shown in Figs. 1.19 and 1.20) and using Eqs. (1.12), (1.13) for optimization of the rotating magnetomotive force, the real relative values of coil turn numbers N^*_{2ri} of the discussed sinusoidal winding are calculated (Table 1.17).

Based on parameters of created windings with $q=3$ (shown in Figs. 1.21 and 1.22) and using Eqs. (1.12), (1.13) for optimization of the rotating magnetomotive force, the real relative values of coil turn numbers N^*_{2ri} of the discussed sinusoidal winding are calculated (Table 1.18).

Based on parameters of created windings with $q=4$ (shown in Figs. 1.23 and 1.24) and using Eqs. (1.12), (1.13) for optimization of the rotating magnetomotive force, the real relative values of coil turn numbers N_{2ri}^{*} of the discussed sinusoidal winding are calculated (Table 1.19).

Based on parameters of created windings with $q=5$ (shown in Figs. 1.25 and 1.26) and using Eqs. (1.12), (1.13) for optimization of the rotating magnetomotive force, the real relative values of coil turn numbers N_{2ri}^{*} of the discussed sinusoidal winding are calculated (Table 1.20).

Based on parameters of created windings with $q=6$ (shown in Figs. 1.27 and 1.28) and using Eqs. (1.12), (1.13) for optimization of the rotating magnetomotive force, the real relative values of coil turn numbers N_{2ri}^{*} of the discussed sinusoidal winding are calculated (Table 1.21).

Table 1.17 Real relative values of coil turn number in coil group for short average pitch simple and STW with optimized rotating magnetomotive force with $q=2$

Number of coil in coil group	Winding type	
	Simple (O_{22})	**Sinusoidal (R_{22})**
1	0.250	**0.366**
2	0.250	**0.1340**

Table 1.18 Real relative values of coil turn number in coil group for short average pitch simple and STW with optimized rotating magnetomotive force with $q=3$

Number of coil in coil group	Winding type	
	Simple (O_{23})	**Sinusoidal (R_{23})**
1	0.1667	**0.266**
2	0.1667	**0.1737**
3	0.1667	**0.0603**

Table 1.19 Real relative values of coil turn number in coil group for short average pitch simple and STW with optimized rotating magnetomotive force with $q=4$

Number of coil in coil group	Winding type	
	Simple (O_{24})	**Sinusoidal (R_{24})**
1	0.1250	**0.207**
2	0.1250	**0.1589**
3	0.1250	**0.0999**
4	0.1250	**0.0341**

Table 1.20 Real relative values of coil turn number in coil group for short average pitch simple and STW with optimized rotating magnetomotive force with $q=5$

Number of coil in coil group	Winding type	
	Simple (O_{25})	Sinusoidal (R_{25})
1	0.10	**0.1691**
2	0.10	**0.1399**
3	0.10	**0.1045**
4	0.10	**0.0646**
5	0.10	**0.0219**

Table 1.21 Real relative values of coil turn number in coil group for short average pitch simple and STW with optimized rotating magnetomotive force with $q=6$

Number of coil in coil group	Winding type	
	Simple (O_{26})	Sinusoidal (R_{26})
1	0.0833	**0.1428**
2	0.0833	**0.1233**
3	0.0833	**0.1000**
4	0.0833	**0.0737**
5	0.0833	**0.0451**
6	0.0833	**0.0152**

1.5   Conclusions

- In sinusoidal three-phase windings, due to unequal number of turns in winding coils, positive and negative half-periods of pulsating magnetomotive force waves of each phase can be created only by separate groups of such coils, which in all cases have to be symmetric in respect of the axes of these groups.
- Sinusoidal three-phase windings can be created from analogical double-layer concentric windings only, because windings of this type completely match structural conditions required by electrical diagram of the sinusoidal three-phase winding.
- According to the average span of concentric windings, sinusoidal three-phase windings can be divided into maximum average pitch and short average pitch windings.
- Maximum and short average pitch sinusoidal three-phase winding is formed from analogical double-layer concentric three-phase windings, leaving the same structure of their electrical diagrams but changing the number of turns in coils which form coil groups, so that this number varied according to sinusoidal law from the symmetry axes of these groups or between the symmetry axes of adjacent groups of the same phase.

- The pulsating magnetomotive force is optimized by varying the number of coil turns in sinusoidal three-phase windings according to sinusoidal law starting from the symmetry axes of coil groups, and the rotating magnetomotive force is optimized when this number is varied according to sinusoidal law starting from the symmetry axes located between adjacent coil groups of the same phase. In both of these cases different results are obtained.
- The distribution of number of turns in coils from all four types of sinusoidal three-phase windings is different when the number of coils in their groups stays the same.

Chapter 2
Electromagnetic Parameters of Sinusoidal Three-Phase Windings

The determination of electromagnetic parameters for three-phase windings analyzed in this work is based on the harmonic analysis of instantaneous periodic functions of rotating magnetomotive force generated by these windings. The instantaneous spatial graphical images of such magnetomotive force are created using winding electrical diagram layouts, real relative values of coil turn numbers in coil groups, and phasor diagram of phase currents formed at a respective point of time. The relative values of instantaneous magnitudes of winding currents and their flow directions are identified from this diagram.

Since the rotating magnetomotive forces of the fundamental and higher-order space harmonics generated by three-phase windings do not change their amplitude values over time, it is sufficient to determine instantaneous currents as well as plot the space distribution of magnetomotive force at a single designated time point, for example $t=0$.

It is assumed that the relative values of current amplitudes in phase windings are $I_{mU}^* = I_{mV}^* = I_{mW}^* = 1$. Then the relative values of instantaneous electric current magnitudes in phase windings at time $t=0$ are as follows:

$$\begin{cases} i_U^* = \sin \omega t = \sin 0° = 0; \\ i_V^* = \sin(\omega t - 120°) = \sin(-120°) = -0.866; \\ i_W^* = \sin(\omega t - 240°) = \sin(-240°) = 0.866. \end{cases} \quad (2.1)$$

Based on the determined real relative values of coil turn numbers in coil groups N_i^* (Tables 1.2–1.21) and instantaneous magnitudes of currents in phase windings i^* (2.1), the conditional values of magnetomotive force changes ΔF_n are determined at time $t=0$ in the slots of magnetic circuit:

© Springer International Publishing Switzerland 2016
J.J. Buksnaitis, *Sinusoidal Three-Phase Windings of Electric Machines*,
DOI 10.1007/978-3-319-42931-1_2

$$\Delta F_n = \Delta F_n' \pm \Delta F_n''; \tag{2.2}$$

where $\Delta F_n'$ is the change of magnetic potential difference generated by the side of the coil located in the top layer of n-th slot; $\Delta F_n''$ is the change of magnetic potential difference generated by the side of the coil located in the bottom layer of the same slot.

The space distribution of rotating magnetomotive force at time $t=0$ is created by calculating changes of magnetic potential difference in the slots of magnetic circuit for the corresponding winding. In this way a non-sinusoidal (stair-shaped) periodic space function is obtained.

In general case, a symmetric three-phase current system in the distributed symmetric three-phase winding creates curves of magnetomotive force which move in space and periodically change over time, as well as have a characteristic stair-shaped form (consisting of several rectangular magnetomotive forces of different height and width) (Fig. 2.1).

The conditional magnitudes of ν-th harmonic amplitudes of rotating magnetomotive forces of the analyzed simple and sinusoidal three-phase windings are calculated analytically according to this expression:

$$F_{m\nu} = \frac{4}{\pi \, \nu} \sum_{j=1}^{k} F_{jr} \sin\left(\nu \, \frac{\alpha_j}{2} \right); \tag{2.3}$$

where: k is the number of rectangles forming the half-periods of the stair-shaped magnetomotive force; ν is the number of odd space harmonic; F_{jr} is the conditional height of the j-th rectangle of the stair-shaped magnetomotive force half-period; α_j

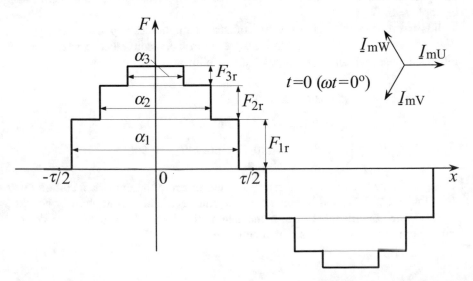

Fig. 2.1 Instantaneous graphical image of the stair-shaped rotating magnetomotive force generated by three-phase windings and symmetric in respect of the coordinate axes

is the width of the j-th rectangle of the stair-shaped magnetomotive force curve, expressed in electrical degrees of the fundamental harmonic.

Since the magnetomotive forces of the higher-order space harmonics negatively impact the operation of the alternating current electrical machines, for this reason the absolute relative values of the amplitudes of these harmonics are considered as negative. All these negative relative magnitudes are combined into a single equivalent magnitude which is equal to the square root of the sum of squares of the relative amplitude values of higher-order magnetomotive force harmonics. Based on such reasoning, three-phase windings of any type can be evaluated from electromagnetic point of view using the electromagnetic efficiency factor, which is expressed as:

$$k_{ef} = 1 - \sqrt{\sum_{v=1}^{\infty} f_v^2 - 1};$$

(2.4)

where f_v is the absolute relative amplitude value of the v-th harmonic of rotating magnetomotive force:

$$f_v = \frac{F_{mv}}{F_{m1}};$$

(2.5)

where F_{m1} is the conditional amplitude value of the first (fundamental) harmonic of rotating magnetomotive force; F_{mv} is the conditional amplitude value of the v-th harmonic of rotating magnetomotive force.

This factor indicates what relative part of rotating magnetomotive force of the fundamental harmonic remains after the negative impact of its higher-order harmonics-induced magnetomotive force is compensated.

From electromagnetic standpoint, the three-phase winding would be optimal if the factor k_{ef} was equal to one. This means that the closer is the value of this indicator to one for a real three-phase winding, the higher is the electromagnetic quality of this winding.

The expression required to calculate the winding factor of the first harmonic for sinusoidal three-phase windings was created based on related literature [12]:

$$k_{w1} = \sum_{i=1}^{q} 2N_i^* \sin\left(\frac{\pi}{2\tau} y_i\right).$$

(2.6)

Similarly, the expression of the winding factor of the v-th harmonic is:

$$k_{wv} = \sum_{i=1}^{q} 2N_i^* \sin\left(v\frac{\pi}{2\tau} y_i\right).$$

(2.7)

Winding factors indicate only the relative values of respective harmonic amplitudes of rotating magnetomotive force in regard of the corresponding harmonic amplitudes of concentrated full-pitched three-phase winding, but do not represent similar reciprocal relative values of harmonics of rotating magnetomotive force for the analyzed three-phase winding. Furthermore, because winding factors form a multi-value system, it is difficult to use these factors to compare three-phase windings of different parameters or windings which belong to several different electromagnetic types.

2.1 Electromagnetic Parameters of Simple and Sinusoidal Three-Phase Windings with $q = 2$

To calculate the conditional magnitudes ΔF_n related to the changes of magnetic potential difference in the slots of magnetic circuit in simple and sinusoidal three-phase windings with $q = 2$, the electrical diagram layouts of these windings presented in Figs. 1.4 and 1.5, earlier-acquired results related to the relative values of coil turn numbers listed in Tables 1.2, 1.7, 1.12, 1.17, as well as the relative values of electric current magnitudes of phase windings determined at time $t = 0$ using equation system (2.1) were used. Values of ΔF_n are calculated using formula (2.2). Calculation results for the discussed windings are listed in Table 2.1.

Symbols from the second row of the table above have the following designations: O_{12}—maximum average pitch concentric (simple) three-phase winding; P_{12}—maximum average pitch STW with the optimized pulsating magnetomotive force; R_{12}—maximum average pitch STW with the optimized rotating magnetomotive force; O_{22}—short average pitch concentric (simple) three-phase winding; P_{22}—short average pitch STW with the optimized pulsating magnetomotive force; R_{22}—sort average pitch STW with the optimized rotating magnetomotive force.

According to the results presented in Table 2.1, the space distributions of magnetomotive force were created for simple and sinusoidal three-phase windings at the selected point in time (Figs. 2.2b and 2.3b).

Table 2.1 Conditional magnitudes related to the changes of magnetic potential difference in the slots of magnetic circuit (ΔF_n) in simple and STW with $q = 2$ at time $t = 0$

Slot no.	Winding type					
	O_{12}	P_{12}	R_{12}	O_{22}	P_{22}	R_{22}
1	0	0	0	−0.2165	−0.204	−0.1160
2	−0.2165	−0.201	−0.232	−0.2165	−0.229	−0.317
3	−0.433	−0.464	−0.402	−0.433	−0.433	−0.433
4	−0.433	−0.402	−0.464	−0.433	−0.433	−0.433
5	−0.433	−0.464	−0.402	−0.2165	−0.229	−0.317
6	−0.2165	−0.201	−0.232	−0.2165	−0.204	−0.1160
7	0	0	0	0.2165	0.204	0.1160
8	0.2165	0.201	0.232	0.2165	0.229	0.317
9	0.433	0.464	0.402	0.433	0.433	0.433
10	0.433	0.402	0.464	0.433	0.433	0.433
11	0.433	0.464	0.402	0.2165	0.229	0.317
12	0.2165	0.201	0.232	0.2165	0.204	0.1160

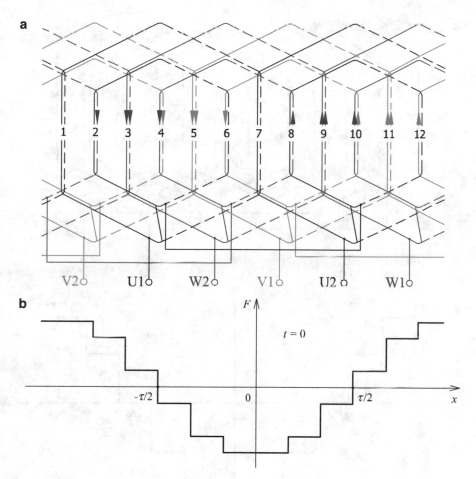

Fig. 2.2 Electrical diagram layout of the maximum average pitch concentric three-phase winding (O_{12}) with $q = 2$ (**a**) and the distribution of its rotating magnetomotive force at $t = 0$ (**b**)

The magnetomotive force space distributions for the other maximum and short average pitch three-phase windings (P_{12}, R_{12}, O_{22}, P_{22}) are similar to those presented above. These distributions differ only in the conditional heights of the magnetomotive force rectangles F_{jr}.

Based on the results from Table 2.1 and figures presented above, the parameters of the negative half-period of rotating magnetomotive forces, which are listed in Table 2.2, were determined.

According to the results calculated using expression (2.3) and presented in Table 2.2, the harmonic analysis of the discussed windings was performed. The results of this analysis are shown in Table 2.3.

Based on the results presented in Table 2.3, the absolute relative values of v-th harmonic amplitudes of rotating magnetomotive forces f_v were calculated for the analyzed windings using expression (2.5) (Table 2.4).

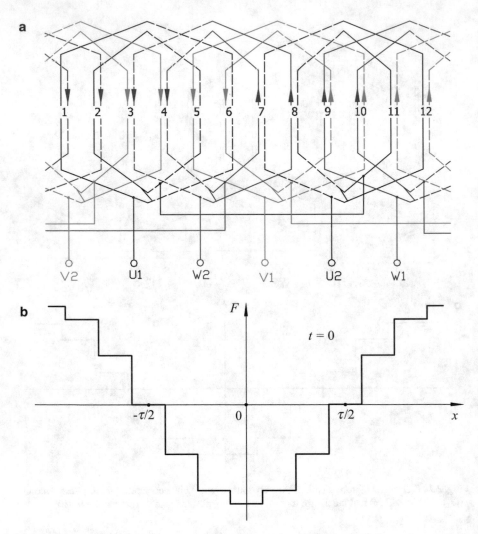

Fig. 2.3 Electrical diagram layout of the short average pitch sinusoidal three-phase winding ($\mathbf{R_{22}}$) with $q = 2$ (**a**) and the distribution of its rotating magnetomotive force at $t = 0$ (**b**)

The electromagnetic efficiency factors k_{ef} of the discussed windings (Table 2.5) were calculated on the basis of results presented in Table 2.4, using expression (2.4). These factors were determined for each winding using the relative amplitude values of rotating magnetomotive forces up to 97-th space harmonic.

Using parameters of the analyzed windings, the winding factors of the first and higher harmonics were calculated for these windings according to formulas (2.6) and (2.7) (Table 2.6).

Table 2.2 Parameters of the negative half-period of rotating magnetomotive forces for simple and STW with $q=2$

Parameter	Winding type					
	O_{12}	P_{12}	R_{12}	O_{22}	P_{22}	R_{22}
k	3	3	3	3	3	3
F_{1r}	−0.2165	−0.201	−0.232	−0.433	−0.433	−0.433
F_{2r}	−0.433	−0.464	−0.402	−0.2165	−0.229	−0.317
F_{3r}	−0.2165	−0.201	−0.232	−0.2165	−0.204	−0.1160
α_1	180°	180°	180°	150°	150°	150°
α_2	120°	120°	120°	90°	90°	90°
α_3	60°	60°	60°	30°	30°	30°

Table 2.3 Harmonic analysis results of rotating magnetomotive forces of simple and STW with $q=2$

ν — Harmonic sequence number	Winding type					
	O_{12}	P_{12}	R_{12}	O_{22}	P_{22}	R_{22}
1	−0.891	−0.896	−0.886	−0.799	−0.806	−0.856
5	0.013	0.026	0	−0.043	−0.037	0
7	−0.009	0.018	0	−0.031	−0.027	0
11	0.081	0.081	0.081	−0.073	−0.073	−0.078
13	−0.069	−0.069	−0.068	0.061	0.062	0.066
17	0.004	0.008	0	0.013	0.011	0
19	−0.003	−0.007	0	0.011	0.010	0
23	0.039	0.039	0.039	0.035	0.035	0.037
25	−0.036	−0.036	−0.035	−0.032	−0.032	−0.034
29	0.002	0.004	0	−0.007	−0.006	0
31	−0.002	−0.004	0	−0.007	−0.006	0

Table 2.4 Absolute relative values of ν-th harmonic amplitudes of rotating magnetomotive forces (f_ν) for simple and STW with $q=2$

ν — Harmonic sequence number	Winding type					
	O_{12}	P_{12}	R_{12}	O_{22}	P_{22}	R_{22}
1	1	1	1	1	1	1
5	0.015	0.029	0	0.054	0.046	0
7	0.010	0.020	0	0.039	0.033	0
11	0.091	0.090	0.091	0.091	0.091	0.091
13	0.077	0.077	0.077	0.076	0.077	0.077
17	0.004	0.009	0	0.016	0.014	0
19	0.003	0.008	0	0.014	0.012	0
23	0.044	0.044	0.044	0.044	0.043	0.043
25	0.040	0.040	0.040	0.040	0.040	0.040
29	0.002	0.004	0	0.009	0.007	0
31	0.002	0.004	0	0.009	0.007	0

Table 2.5 Electromagnetic efficiency factors k_{ef} of simple and STW with $q=2$

Winding type					
O_{12}	P_{12}	R_{12}	O_{22}	P_{22}	R_{22}
0.8517	0.8485	0.8531	0.8364	0.8412	0.8534

Table 2.6 Winding factors of the first and higher harmonics ($k_{w\,\nu}$) of simple and STW with $q=2$

ν—Harmonic sequence number	Winding type					
	O_{12}	P_{12}	R_{12}	O_{22}	P_{22}	R_{22}
1	0.933	0.938	0.928	0.8365	0.844	0.897
5	0.0670	0.1342	0	−0.224	−0.1971	0
7	−0.0670	−0.1342	0	−0.224	−0.1971	0
11	−0.933	−0.938	−0.928	0.8365	0.844	0.897
13	0.933	0.938	0.928	−0.8365	−0.844	−0.897
17	0.0670	0.1342	0	0.224	0.1971	0
19	−0.0670	−0.1342	0	0.224	0.1971	0

2.2   Electromagnetic Parameters of Simple and Sinusoidal Three-Phase Windings with $q=3$

To calculate the conditional magnitudes ΔF_n related to the changes of magnetic potential difference in the slots of magnetic circuit in simple and sinusoidal three-phase windings with $q=3$, the electrical diagram layouts of these windings presented in Figs. 1.9 and 1.21, earlier-acquired results related to the relative values of coil turn numbers listed in Tables 1.3, 1.8, 1.13, 1.18, as well as the relative values of electric current magnitudes of phase windings determined at time $t=0$ using equation system (2.1) were used. Values of ΔF_n are calculated using formula (2.2). Calculation results for the discussed windings are listed in Table 2.7.

According to the results presented in Table 2.7, the space distributions of magnetomotive force were created for simple and sinusoidal three-phase windings at the selected point in time (Figs. 2.4b and 2.5b).

The magnetomotive force space distributions for the other maximum and short average pitch three-phase windings (P_{13}, R_{13}, O_{23} P_{23}) are similar to those presented above. These distributions differ only in the conditional heights of the magnetomotive force rectangles F_{jr}.

Based on the results from Table 2.7 and figures presented above, the parameters of the negative half-period of rotating magnetomotive forces, which are listed in Table 2.8, were determined.

Table 2.7 Conditional magnitudes related to the changes of magnetic potential difference in the slots of magnetic circuit (ΔF_n) in simple and STW with $q=3$ at time $t=0$

Slot no.	Winding type					
	O_{13}	P_{13}	R_{13}	O_{23}	P_{23}	R_{23}
1	0	0	0	−0.1444	−0.1116	−0.0522
2	−0.1444	−0.1226	−0.1044	−0.1444	−0.1503	−0.1504
3	−0.1444	−0.1503	−0.1963	−0.1444	−0.1710	−0.230
4	−0.289	−0.320	−0.264	−0.289	−0.283	−0.282
5	−0.289	−0.273	−0.301	−0.289	−0.301	−0.301
6	−0.289	−0.273	−0.301	−0.289	−0.283	−0.282
7	−0.289	−0.320	−0.264	−0.1444	−0.1710	−0.230
8	−0.1444	−0.1503	−0.1963	−0.1444	−0.1503	−0.1504
9	−0.1444	−0.1226	−0.1044	−0.1444	−0.1116	−0.0522
10	0	0	0	0.1444	0.1116	0.0522
11	0.1444	0.1226	0.1044	0.1444	0.1503	0.1504
12	0.1444	0.1503	0.1963	0.1444	0.1710	0.230
13	0.289	0.320	0.264	0.289	0.283	0.282
14	0.289	0.273	0.301	0.289	0.301	0.301
15	0.289	0.273	0.301	0.289	0.283	0.282
16	0.289	0.320	0.264	0.1444	0.1710	0.230
17	0.1444	0.1503	0.1963	0.1444	0.1503	0.1504
18	0.1444	0.1226	0.1044	0.1444	0.1116	0.0522

According to the results calculated using expression (2.3) and presented in Table 2.8, the harmonic analysis of the discussed windings was performed. The results of this analysis are shown in Table 2.9.

Based on the results presented in Table 2.9, the absolute relative values of ν-th harmonic amplitudes of rotating magnetomotive forces f_ν were calculated for the analyzed windings using expression (2.5) (Table 2.10).

The electromagnetic efficiency factors k_{ef} of the discussed windings (Table 2.11) were calculated on the basis of results presented in Table 2.10, using expression (2.4). These factors were determined for each winding using the relative amplitude values of rotating magnetomotive forces up to 97-th space harmonic.

Using parameters of the analyzed windings, the winding factors of the first and higher harmonics were calculated for these windings according to formulas (2.6) and (2.7) (Table 2.12).

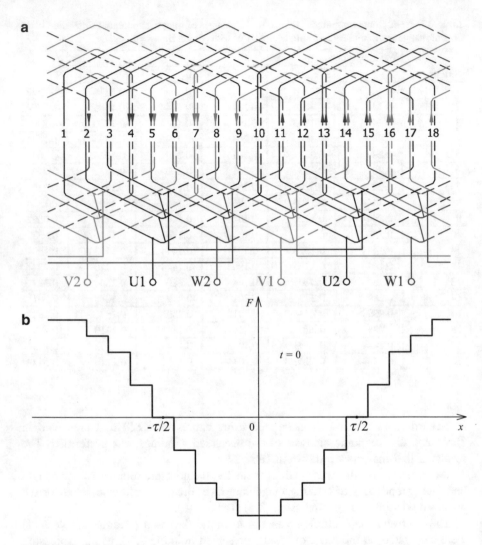

Fig. 2.4 Electrical diagram layout of the maximum average pitch concentric three-phase winding ($\mathbf{O_{13}}$) with $q = 3$ (**a**) and the distribution of its rotating magnetomotive force at $t = 0$ (**b**)

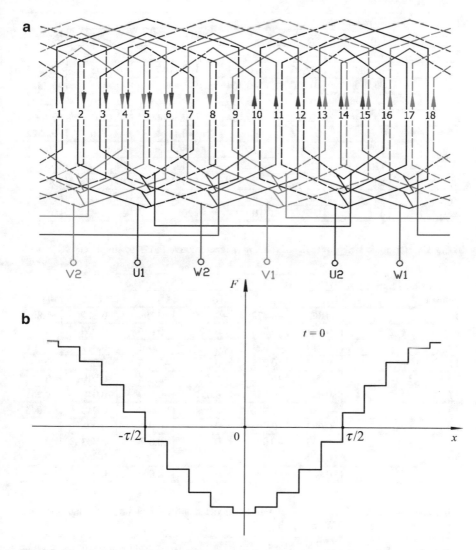

Fig. 2.5 Electrical diagram layout of the short average pitch sinusoidal three-phase winding (R_{23}) with $q = 3$ (**a**) and the distribution of its rotating magnetomotive force at $t = 0$ (**b**)

Table 2.8 Parameters of the negative half-period of rotating magnetomotive forces for simple and STW with $q=3$

Parameter	Winding type					
	O_{13}	P_{13}	R_{13}	O_{23}	P_{23}	R_{23}
k	4	4	4	5	5	5
F_{1r}	−0.289	−0.273	−0.301	−0.1445	−0.1505	−0.1505
F_{2r}	−0.289	−0.320	−0.264	−0.289	−0.283	−0.282
F_{3r}	−0.1444	−0.1503	−0.1963	−0.1444	−0.1710	−0.230
F_{4r}	−0.1444	−0.1226	−0.1044	−0.1444	−0.1503	−0.1504
F_{5r}	0	0	0	−0.1444	−0.1116	−0.0522
α_1	160°	160°	160°	180°	180°	180°
α_2	120°	120°	120°	140°	140°	140°
α_3	80°	80°	80°	100°	100°	100°
α_4	40°	40°	40°	60°	60°	60°
α_5	0	0	0	20°	20°	20°

Table 2.9 Harmonic analysis results of rotating magnetomotive forces of simple and STW with $q=3$

ν–Harmonic sequence number	Winding type					
	O_{13}	P_{13}	R_{13}	O_{23}	P_{23}	R_{23}
1	−0.862	−0.872	−0.875	−0.794	−0.817	−0.861
5	−0.007	0.008	0	−0.036	−0.026	0
7	−0.019	−0.021	0	−0.021	−0.012	0
11	0.012	0.013	0	−0.013	−0.008	0
13	0.003	−0.003	0	−0.014	−0.010	0
17	0.051	0.051	0.051	−0.047	−0.048	−0.051
19	−0.045	−0.046	−0.046	0.042	0.043	0.045
23	−0.002	0.002	0	0.008	0.006	0
25	−0.005	−0.006	0	0.006	0.003	0
29	0.004	0.005	0	0.005	0.003	0
31	0.001	−0.001	0	0.006	0.004	0

Table 2.10 Absolute relative values of ν-th harmonic amplitudes of rotating magnetomotive forces (f_ν) for simple and STW with $q=3$

ν–Harmonic sequence number	Winding type					
	O_{13}	P_{13}	R_{13}	O_{23}	P_{23}	R_{23}
1	1	1	1	1	1	1
5	0.008	0.009	0	0.045	0.032	0
7	0.022	0.024	0	0.026	0.015	0
11	0.014	0.015	0	0.016	0.010	0
13	0.003	0.003	0	0.018	0.012	0
17	0.059	0.058	0.058	0.059	0.059	0.059
19	0.052	0.053	0.053	0.053	0.053	0.052
23	0.002	0.002	0	0.010	0.007	0
25	0.006	0.007	0	0.008	0.004	0
29	0.005	0.006	0	0.006	0.004	0
31	0.001	0.001	0	0.008	0.005	0

Table 2.11 Electromagnetic efficiency factors k_{ef} of simple and STW with $q=3$

Winding type					
O_{13}	P_{13}	R_{13}	O_{23}	P_{23}	R_{23}
0.9001	0.8993	0.9047	0.8864	0.8965	0.9045

Table 2.12 Winding factors of the first and higher harmonics ($k_{w\,\nu}$) of simple and STW with $q=3$

ν—Harmonic sequence number	Winding type					
	O_{13}	P_{13}	R_{13}	O_{23}	P_{23}	R_{23}
1	0.902	0.913	0.916	0.831	0.855	0.902
5	−0.0378	0.0432	0	−0.1884	−0.1350	0
7	−0.1359	−0.1528	0	−0.1536	−0.0883	0
11	−0.1359	−0.1528	0	0.1536	0.0883	0
13	−0.0378	0.0432	0	0.1884	0.1350	0
17	0.902	0.913	0.916	−0.831	−0.855	−0.902
19	−0.902	−0.913	−0.916	0.831	0.855	0.902

2.3   Electromagnetic Parameters of Simple and Sinusoidal Three-Phase Windings with $q=4$

To calculate the conditional magnitudes ΔF_n related to the changes of magnetic potential difference in the slots of magnetic circuit in simple and sinusoidal three-phase windings with $q=4$, the electrical diagram layouts of these windings presented in Figs. 1.11 and 1.23, earlier-acquired results related to the relative values of coil turn numbers listed in Tables 1.4, 1.9, 1.14, 1.19, as well as the relative values of electric current magnitudes of phase windings determined at time $t=0$ using equation system (2.1) were used. Values of ΔF_n are calculated using formula (2.2). Calculation results for the discussed windings are listed in Table 2.13.

According to the results presented in Table 2.13, the space distributions of magnetomotive force were created for simple and sinusoidal three-phase windings at the selected point in time (Figs. 2.6b and 2.7b).

The magnetomotive force space distributions for the other maximum and short average pitch three-phase windings (P_{14}, R_{14}, O_{24}, P_{24}) are similar to those presented above. These distributions differ only in the conditional heights of the magnetomotive force rectangles F_{jr}.

Based on the results from Table 2.13 and figures presented above, the parameters of the negative half-period of rotating magnetomotive forces, which are listed in Table 2.14, were determined.

According to the results calculated using expression (2.3) and presented in Table 2.14, the harmonic analysis of the discussed windings was performed. The results of this analysis are shown in Table 2.15.

Based on the results presented in Table 2.15, the absolute relative values of v-th harmonic amplitudes of rotating magnetomotive forces f_v were calculated for the analyzed windings using expression (2.5) (Table 2.16).

The electromagnetic efficiency factors k_{ef} of the discussed windings (Table 2.17) were calculated on the basis of results presented in Table 2.16, using expression (2.4). These factors were determined for each winding using the relative amplitude values of rotating magnetomotive forces up to 97-th space harmonic.

Using parameters of the analyzed windings, the winding factors of the first and higher harmonics were calculated for these windings according to formulas (2.6) and (2.7) (Table 2.18).

Table 2.13 Conditional magnitudes related to the changes of magnetic potential difference in the slots of magnetic circuit (ΔF_n) in simple and STW with $q=4$ at time $t=0$

Slot no.	Winding type					
	O_{14}	P_{14}	R_{14}	O_{24}	P_{24}	R_{24}
1	0	0	0	–0.10825	–0.0795	–0.0295
2	–0.10825	–0.0865	–0.0590	–0.10825	–0.1036	–0.0865
3	–0.10825	–0.1059	–0.1141	–0.10825	–0.1205	–0.1376
4	–0.10825	–0.1182	–0.1612	–0.10825	–0.1294	–0.1793
5	–0.2165	–0.245	–0.1974	–0.2165	–0.209	–0.209
6	–0.2165	–0.205	–0.220	–0.2165	–0.224	–0.224
7	–0.2165	–0.212	–0.228	–0.2165	–0.224	–0.224
8	–0.2165	–0.205	–0.220	–0.2165	–0.209	–0.209
9	–0.2165	–0.245	–0.1974	–0.10825	–0.1294	–0.1793
10	–0.10825	–0.1182	–0.1612	–0.10825	–0.1205	–0.1376
11	–0.10825	–0.1059	–0.1141	–0.10825	–0.1036	–0.0865
12	–0.10825	–0.0865	–0.0590	–0.10825	–0.0795	–0.0295
13	0	0	0	0.10825	0.0795	0.0295
14	0.10825	0.0865	0.0590	0.10825	0.1036	0.0865
15	0.10825	0.1059	0.1141	0.10825	0.1205	0.1376
16	0.10825	0.1182	0.1612	0.10825	0.1294	0.1793
17	0.2165	0.245	0.1974	0.2165	0.209	0.209
18	0.2165	0.205	0.220	0.2165	0.224	0.224
19	0.2165	0.212	0.228	0.2165	0.224	0.224
20	0.2165	0.205	0.220	0.2165	0.209	0.209
21	0.2165	0.245	0.1974	0.10825	0.1294	0.1793
22	0.10825	0.1182	0.1612	0.10825	0.1205	0.1376
23	0.10825	0.1059	0.1141	0.10825	0.1036	0.0865
24	0.10825	0.0865	0.0590	0.10825	0.0795	0.0295

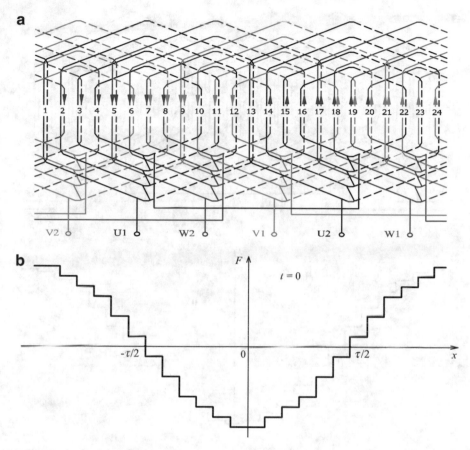

Fig. 2.6 Electrical diagram layout of the maximum average pitch concentric three-phase winding (O_{14}) with $q = 4$ (**a**) and the distribution of its rotating magnetomotive force at $t = 0$ (**b**)

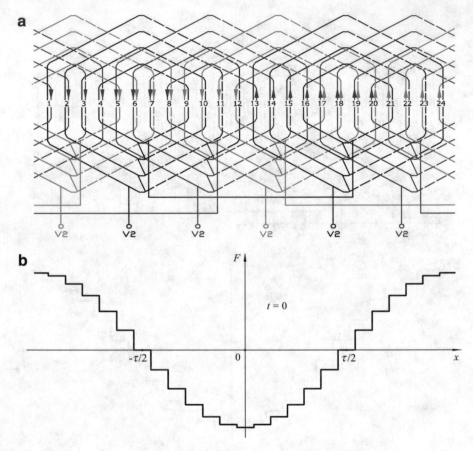

Fig. 2.7 Electrical diagram layout of the short average pitch sinusoidal three-phase winding ($\mathbf{R}_{24}$) with $q = 4$ (**a**) and the distribution of its rotating magnetomotive force at $t = 0$ (**b**)

Table 2.14 Parameters of the negative half-period of rotating magnetomotive forces for simple and STW with $q=4$

	Winding type					
Parameter	O_{14}	P_{14}	R_{14}	O_{24}	P_{24}	R_{24}
k	6	6	6	6	6	6
F_{1r}	−0.10825	−0.106	−0.1140	−0.2165	−0.224	−0.224
F_{2r}	−0.2165	−0.205	−0.220	−0.2165	−0.209	−0.209
F_{3r}	−0.2165	−0.245	−0.1974·	−0.10825	−0.1294	−0.1793
F_{4r}	−0.10825	−0.1182	−0.1612	−0.10825	−0.1205	−0.1376
F_{5r}	−0.10825	−0.1059	−0.1141	−0.10825	−0.1036	−0.0865
F_{6r}	−0.10825	−0.0865	−0.0590	−0.10825	−0.0795	−0.0295
α_1	180°	180°	180°	165°	165°	165°
α_2	150°	150°	150°	135°	135°	135°
α_3	120°	120°	120°	105°	105°	105°
α_4	90°	90°	90°	75°	75°	75°
α_5	60°	60°	60°	45°	45°	45°
α_6	30°	30°	30°	15°	15°	15°

Table 2.15 Harmonic analysis results of rotating magnetomotive forces of simple and STW with $q=4$

ν–Harmonic sequence number	Winding type					
	O_{14}	P_{14}	R_{14}	O_{24}	P_{24}	R_{24}
1	−0.845	−0.860	−0.871	−0.792	−0.816	−0.863
5	−0.015	0	0	−0.034	−0.025	0
7	−0.020	−0.019	0	−0.019	−0.010	0
11	0.004	0.008	0	−0.009	−0.006	0
13	−0.004	−0.007	0	−0.008	−0.005	0
17	0.008	0.008	0	−0.008	−0.004	0
19	0.004	0	0	−0.009	−0.007	0
23	0.037	0.037	0.038	−0.034	−0.035	−0.038
25	−0.034	−0.034	−0.035	0.032	−0.033	−0.035
29	−0.003	0	0	0.006	−0.004	0
31	−0.004	−0.004	0	0.004	−0.002	0

Table 2.16 Absolute relative values of v-th harmonic amplitudes of rotating magnetomotive forces (f_v) for simple and STW with $q=4$

v–Harmonic sequence number	Winding type					
	O_{14}	P_{14}	R_{14}	O_{24}	P_{24}	R_{24}
1	1	1	1	1	1	1
5	0.018	0	0	0.043	0.031	0
7	0.024	0.022	0	0.024	0.012	0
11	0.005	0.009	0	0.011	0.007	0
13	0.005	0.008	0	0.010	0.006	0
17	0.009	0.009	0	0.010	0.005	0
19	0.005	0	0	0.011	0.009	0
23	0.044	0.043	0.44	0.043	0.043	0.044
25	0.040	0.040	0.40	0.040	0.040	0.041
29	0.004	0	0	0.008	0.005	0
31	0.005	0.005	0	0.005	0.002	0

Table 2.17 Electromagnetic efficiency factors k_{ef} of simple and STW with $q=4$

Winding type					
O_{14}	P_{14}	R_{14}	O_{24}	P_{24}	R_{24}
0.9216	0.9245	0.9292	0.9103	0.9207	0.9289

Table 2.18 Winding factors of the first and higher harmonics (k_{wv}) of simple and STW with $q=4$

v–Harmonic sequence number	Winding type					
	O_{14}	P_{14}	R_{14}	O_{24}	P_{24}	R_{24}
1	0.885	0.899	0.912	0.829	0.855	0.904
5	−0.0786	0.00015	0	−0.1778	−0.1306	0
7	−0.1456	−0.1414	0	−0.1365	−0.0755	0
11	−0.0482	−0.0895	0	0.1092	0.0676	0
13	0.0482	0.0895	0	0.1092	0.0676	0
17	0.1456	0.1414	0	−0.1365	−0.755	0
19	0.0786	−0.00015	0	−0.1778	−0.1306	0
23	−0.885	−0.899	−0.912	0.829	0.855	0.904
25	0.885	0.899	0.912	−0.829	−0.855	−0.904

2.4   Electromagnetic Parameters of Simple and Sinusoidal Three-Phase Windings with $q=5$

To calculate the conditional magnitudes ΔF_n related to the changes of magnetic potential difference in the slots of magnetic circuit in simple and sinusoidal three-phase windings with $q=5$, the electrical diagram layouts of these windings presented in Figs. 1.13 and 1.25, earlier-acquired results related to the relative values of coil turn numbers listed in Tables 1.5, 1.10, 1.15, 1.20, as well as the relative values of electric current magnitudes of phase windings determined at time $t=0$ using equation system (2.1) were used. Values of ΔF_n are calculated using formula (2.2). Calculation results for the discussed windings are listed in Table 2.19.

According to the results presented in Table 2.19, the space distributions of magnetomotive force were created for simple and sinusoidal three-phase windings at the selected point in time (Figs. 2.8b and 2.9b).

The magnetomotive force space distributions for the other maximum and short average pitch three-phase windings (P_{15}, R_{15}, O_{25}, P_{25}) are similar to those presented above. These distributions differ only in the conditional heights of the magnetomotive force rectangles F_{jr}.

Based on the results from Table 2.19 and figures presented above, the parameters of the negative half-period of rotating magnetomotive forces, which are listed in Table 2.20, were determined.

According to the results calculated using expression (2.3) and presented in Table 2.20, the harmonic analysis of the discussed windings was performed. The results of this analysis are shown in Table 2.21.

Based on the results presented in Table 2.21, the absolute relative values of v-th harmonic amplitudes of rotating magnetomotive forces f_v were calculated for the analyzed windings using expression (2.5) (Table 2.22).

The electromagnetic efficiency factors k_{ef} of the discussed windings (Table 2.23) were calculated on the basis of results presented in Table 2.22, using expression (2.4). These factors were determined for each winding using the relative amplitude values of rotating magnetomotive forces up to 97-th space harmonic.

Using parameters of the analyzed windings, the winding factors of the first and higher harmonics were calculated for these windings according to formulas (2.6) and (2.7) (Table 2.24).

Table 2.19 Conditional magnitudes related to the changes of magnetic potential difference in the slots of magnetic circuit (ΔF_n) in simple and STW with $q=5$ at time $t=0$

Slot no.	Winding type					
	O_{15}	P_{15}	R_{15}	O_{25}	P_{25}	R_{25}
1	0	0	0	−0.0866	−0.0615	−0.01897
2	−0.0866	−0.0663	−0.0378	−0.0866	−0.0777	−0.0559
3	−0.0866	−0.0802	−0.0740	−0.0866	−0.0905	−0.0905
4	−0.0866	−0.0905	−0.1070	−0.0866	−0.0994	−0.1212
5	−0.0866	−0.0969	−0.1353	−0.0866	−0.1039	−0.1464
6	−0.1732	−0.1981	−0.1576	−0.1732	−0.1654	−0.1654
7	−0.1732	−0.1632	−0.1731	−0.1732	−0.1771	−0.1771
8	−0.1732	−0.1707	−0.1811	−0.1732	−0.1810	−0.1810
9	−0.1732	−0.1707	−0.1811	−0.1732	−0.1771	−0.1771
10	−0.1732	−0.1632	−0.1731	−0.1732	−0.1654	−0.1654
11	−0.1732	−0.1981	−0.1576	−0.0866	−0.1039	−0.1464
12	−0.0866	−0.0969	−0.1353	−0.0866	−0.0994	−0.1212
13	−0.0866	−0.0905	−0.1070	−0.0866	−0.0905	−0.0905
14	−0.0866	−0.0802	−0.0740	−0.0866	−0.0777	−0.0559
15	−0.0866	−0.0663	−0.0378	−0.0866	−0.0615	−0.01897
16	0	0	0	0.0866	0.0615	0.01897
17	0.0866	0.0663	0.0378	0.0866	0.0777	0.0559
18	0.0866	0.0802	0.0740	0.0866	0.0905	0.0905
19	0.0866	0.0905	0.1070	0.0866	0.0994	0.1212
20	0.0866	0.0969	0.1353	0.0866	0.1039	0.1464
21	0.1732	0.1981	0.1576	0.1732	0.1654	0.1654
22	0.1732	0.1632	0.1731	0.1732	0.1771	0.1771
23	0.1732	0.1707	0.1811	0.1732	0.1810	0.1810
24	0.1732	0.1707	0.1811	0.1732	0.1771	0.1771
25	0.1732	0.1632	0.1731	0.1732	0.1654	0.1654
26	0.1732	0.1981	0.1576	0.0866	0.1039	0.1464
27	0.0866	0.0969	0.1353	0.0866	0.0994	0.1212
28	0.0866	0.0905	0.1070	0.0866	0.0905	0.0905
29	0.0866	0.0802	0.0740	0.0866	0.0777	0.0559
30	0.0866	0.0663	0.0378	0.0866	0.0615	0.01897

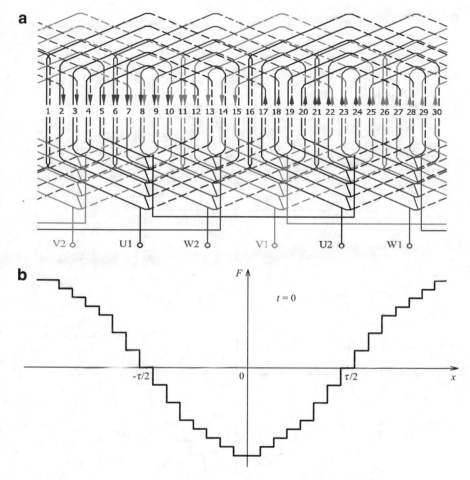

Fig. 2.8 Electrical diagram layout of the maximum average pitch concentric three-phase winding (O_{15}) with $q = 5$ (**a**) and the distribution of its rotating magnetomotive force at $t = 0$ (**b**)

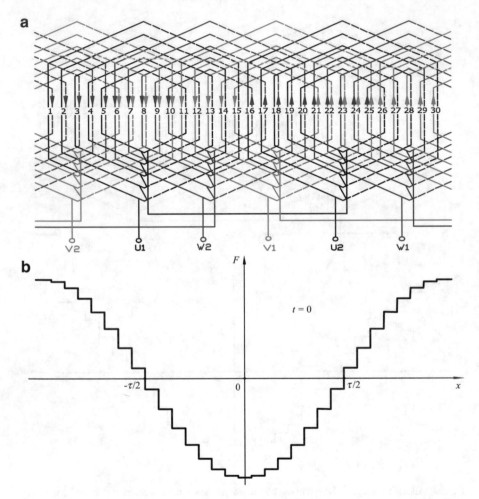

Fig. 2.9 Electrical diagram layout of the short average pitch sinusoidal three-phase winding ($\mathbf{R_{25}}$) with $q = 5$ (**a**) and the distribution of its rotating magnetomotive force at $t = 0$ (**b**)

Table 2.20 Parameters of the negative half-period of rotating magnetomotive forces for simple and STW with $q=5$

Parameter	Winding type					
	O_{15}	P_{15}	R_{15}	O_{25}	P_{25}	R_{25}
k	7	7	7	8	8	8
F_{1r}	−0.1732	−0.1707	−0.1811	−0.0866	−0.0905	−0.0905
F_{2r}	−0.1732	−0.1632	−0.1731	−0.1732	−0.1771	−0.1771
F_{3r}	−0.1732	−0.1981	−0.1576	−0.1732	−0.1654	−0.1654
F_{4r}	−0.0866	−0.0969	−0.1353	−0.0866	−0.1039	−0.1464
F_{5r}	−0.0866	−0.0905	−0.1070	−0.0866	−0.0994	−0.1212
F_{6r}	−0.0866	−0.0802	−0.0740	−0.0866	−0.0905	−0.0905
F_{7r}	−0.0866	−0.0663	−0.0378	−0.0866	−0.0777	−0.0559
F_{8r}	0	0	0	−0.0866	−0.0615	−0.01897
α_1	168°	168°	168°	180°	180°	180°
α_2	144°	144°	144°	156°	156°	156°
α_3	v120°	120°	120°	132°	132°	132°
α_4	96°	96°	96°	108°	108°	108°
α_5	72°	72°	72°	84°	84°	84°
α_6	48°	48°	48°	60°	60°	60°
α_7	24°	24°	24°	36°	36°	36°
α_8	0	0	0	12°	12°	12°

Table 2.21 Harmonic analysis results of rotating magnetomotive forces of simple and STW with $q=5$

ν −Harmonic sequence number	Winding type					
	O_{15}	P_{15}	R_{15}	O_{25}	P_{25}	R_{25}
1	−0.835	−0.851	−0.869	−0.791	−0.816	−0.864
5	−0.019	−0.005	0	−0.033	−0.025	0
7	−0.020	−0.018	0	−0.018	−0.010	0
11	0.001	0.005	0	−0.008	−0.005	0
13	−0.005	−0.007	0	−0.007	−0.004	0
17	0.004	0.005	0	−0.005	−0.003	0
19	−0.001	−0.003	0	−0.005	−0.003	0
23	0.006	0.005	0	−0.005	−0.003	0
25	0.004	0.001	0	−0.007	−0.005	0
29	0.029	0.029	0.030	−0.027	−0.028	−0.030
31	−0.027	−0.027	−0.028	0.026	0.026	0.028
35	−0.003	−0.001	0	0.005	0.004	0

Table 2.22 Absolute relative values of v-th harmonic amplitudes of rotating magnetomotive forces (f_v) for simple and STW with $q=5$

v—Harmonic sequence number	Winding type					
	O_{15}	P_{15}	R_{15}	O_{25}	P_{25}	R_{25}
1	1	1	1	1	1	1
5	0.023	0.006	0	0.042	0.031	0
7	0.024	0.021	0	0.023	0.012	0
11	0.001	0.006	0	0.010	0.006	0
13	0.006	0.008	0	0.009	0.005	0
17	0.005	0.006	0	0.006	0.004	0
19	0.001	0.004	0	0.006	0.004	0
23	0.007	0.006	0	0.006	0.004	0
25	0.005	0.001	0	0.009	0.006	0
29	0.035	0.034	0.035	0.034	0.034	0.035
31	0.032	0.032	0.032	0.033	0.032	0.032
35	0.004	0.001	0	0.006	0.005	0

Table 2.23 Electromagnetic efficiency factors k_{ef} of simple and STW with $q=5$

Winding type					
O_{15}	P_{15}	R_{15}	O_{25}	P_{25}	R_{25}
0.9340	0.9394	0.9448	0.9237	0.9346	0.9447

Table 2.24 Winding factors of the first and higher harmonics ($k_{w\,v}$) of simple and STW with $q=5$

v—Harmonic sequence number	Winding type					
	O_{15}	P_{15}	R_{15}	O_{25}	P_{25}	R_{25}
1	0.874	0.891	0.910	0.829	0.855	0.905
5	−0.100	−0.0256	0	−0.1732	−0.1285	0
7	−0.1462	−0.1305	0	−0.1294	−0.0704	0
11	−0.01144	−0.0572	0	0.0948	0.0605	0
13	0.0684	0.0910	0	0.0885	0.0526	0
17	0.0684	0.0910	0	−0.0885	−0.0526	0
19	−0.01144	−0.0572	0	−0.0948	−0.0605	0
23	−0.1462	−0.1305	0	0.1294	0.0704	0
25	−0.100	−0.0256	0	0.1732	0.1285	0
29	0.874	0.891	0.910	−0.829	−0.855	−0.905
31	−0.874	−0.891	−0.910	0.829	0.855	0.905

2.5   Electromagnetic Parameters of Simple and Sinusoidal Three-Phase Windings with $q=6$

To calculate the conditional magnitudes ΔF_n related to the changes of magnetic potential difference in the slots of magnetic circuit in simple and sinusoidal three-phase windings with $q=6$, the electrical diagram layouts of these windings presented in Figs. 1.15 and 1.27, earlier-acquired results related to the relative values of coil turn numbers listed in Tables 1.6, 1.11, 1.16, 1.21, as well as the relative values of electric current magnitudes of phase windings determined at time $t=0$ using equation system (2.1) were used. Values of ΔF_n are calculated using formula (2.2). Calculation results for the discussed windings are listed in Table 2.25.

According to the results presented in Table 2.25, the space distributions of magnetomotive force were created for simple and sinusoidal three-phase windings at the selected point in time (Figs. 2.10b and 2.11b).

The magnetomotive force space distributions for the other maximum and short average pitch three-phase windings (P_{16}, R_{16}, O_{26}, P_{26}) are similar to those presented above. These distributions differ only in the conditional heights of the magnetomotive force rectangles F_{jr}.

Based on the results from Table 2.25 and figures presented above, the parameters of the negative half-period of rotating magnetomotive forces, which are listed in Table 2.26, were determined.

According to the results calculated using expression (2.3) and presented in Table 2.26, the harmonic analysis of the discussed windings was performed. The results of this analysis are shown in Table 2.27.

Based on the results presented in Table 2.27, the absolute relative values of v-th harmonic amplitudes of rotating magnetomotive forces f_v were calculated for the analyzed windings using expression (2.5) (Table 2.28).

The electromagnetic efficiency factors k_{ef} of the discussed windings (Table 2.29) were calculated on the basis of results presented in Table 2.28, using expression (2.4). These factors were determined for each winding using the relative amplitude values of rotating magnetomotive forces up to 97-th space harmonic.

Using parameters of the analyzed windings, the winding factors of the first and higher harmonics were calculated for these windings according to formulas (2.6) and (2.7) (Table 2.30).

Table 2.25 Conditional magnitudes related to the changes of magnetic potential difference in the slots of magnetic circuit (ΔF_n) in simple and STW with $q=6$ at time $t=0$

Slot no.	Winding type					
	O_{16}	P_{16}	R_{16}	O_{26}	P_{26}	R_{26}
1	0	0	0	−0.0722	−0.0500	−0.01316
2	−0.0722	−0.0535	−0.0263	−0.0722	−0.0617	−0.0391
3	−0.0722	−0.0638	−0.0518	−0.0722	−0.0714	−0.0638
4	−0.0722	−0.0721	−0.0758	−0.0722	−0.0790	−0.0866
5	−0.0722	−0.0783	−0.0974	−0.0722	−0.0842	−0.1068
6	−0.0722	−0.0820	−0.1160	−0.0722	−0.0869	−0.1237
7	−0.1443	−0.1666	−0.1313	−0.1443	−0.1368	−0.1368
8	−0.1443	−0.1355	−0.1424	−0.1443	−0.1458	−0.1458
9	−0.1443	−0.1421	−0.1492	−0.1443	−0.1503	−0.1504
10	−0.1443	−0.1443	−0.1515	−0.1443	−0.1503	−0.1504
11	−0.1443	−0.1421	−0.1492	−0.1443	−0.1458	−0.1458
12	−0.1443	−0.1355	−0.1424	−0.1443	−0.1368	−0.1368
13	−0.1443	−0.1666	−0.1313	−0.0722	−0.0869	−0.1237
14	−0.0722	−0.0820	−0.1160	−0.0722	−0.0842	−0.1068
15	−0.0722	−0.0783	−0.0974	−0.0722	−0.0790	−0.0866
16	−0.0722	−0.0721	−0.0758	−0.0722	−0.0714	−0.0638
17	−0.0722	−0.0638	−0.0518	−0.0722	−0.0617	−0.0391
18	−0.0722	−0.0535	−0.0263	−0.0722	−0.0500	−0.01316
19	0	0	0	0.0722	0.0500	0.01316
20	0.0722	0.0535	0.0263	0.0722	0.0617	0.0391
21	0.0722	0.0638	0.0518	0.0722	0.0714	0.0638
22	0.0722	0.0721	0.0758	0.0722	0.0790	0.0866
23	0.0722	0.0783	0.0974	0.0722	0.0842	0.1068
24	0.0722	0.0820	0.1160	0.0722	0.0869	0.1237
25	0.1443	0.1666	0.1313	0.1443	0.1368	0.1368
26	0.1443	0.1355	0.1424	0.1443	0.1458	0.1458
27	0.1443	0.1421	0.1492	0.1443	0.1503	0.1504
28	0.1443	0.1443	0.1515	0.1443	0.1503	0.1504
29	0.1443	0.1421	0.1492	0.1443	0.1458	0.1458
30	0.1443	0.1355	0.1424	0.1443	0.1368	0.1368
31	0.1443	0.1666	0.1313	0.0722	0.0869	0.1237
32	0.0722	0.0820	0.1160	0.0722	0.0842	0.1068
33	0.0722	0.0783	0.0974	0.0722	0.0790	0.0866
34	0.0722	0.0721	0.0758	0.0722	0.0714	0.0638
35	0.0722	0.0638	0.0518	0.0722	0.0617	0.0391
36	0.0722	0.0535	0.0263	0.0722	0.0500	0.01316

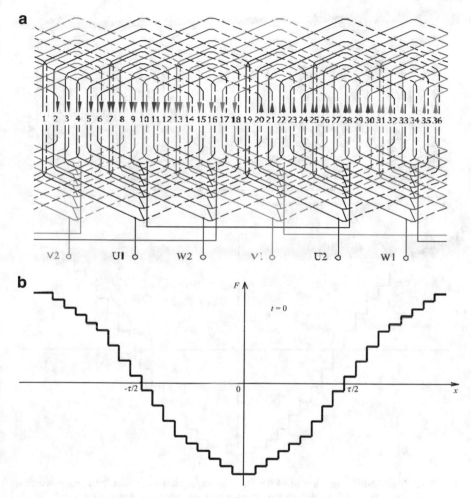

Fig. 2.10 Electrical diagram layout of the maximum average pitch concentric three-phase winding (O_{16}) with $q = 6$ (**a**) and the distribution of its rotating magnetomotive force at $t = 0$ (**b**)

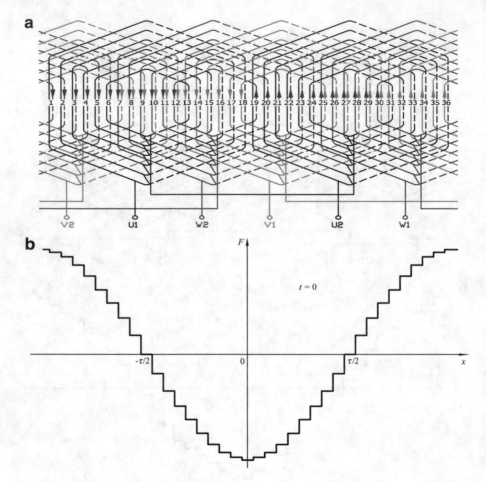

Fig. 2.11 Electrical diagram layout of the short average pitch sinusoidal three-phase winding ($\mathbf{R_{26}}$) with $q = 6$ (**a**) and the distribution of its rotating magnetomotive force at $t = 0$ (**b**)

Table 2.26 Parameters of the negative half-period of rotating magnetomotive forces for simple and STW with $q=6$

Parameter	Winding type					
	O_{16}	P_{16}	R_{16}	O_{26}	P_{26}	R_{26}
k	9	9	9	9	9	9
F_{1r}	−0.0722	−0.0722	−0.0758	−0.1443	−0.1503	−0.1504
F_{2r}	−0.1443	−0.1421	−0.1492	−0.1443	−0.1458	−0.1458
F_{3r}	−0.1443	−0.1355	−0.1424	−0.1443	−0.1368	−0.1368
F_{4r}	−0.1443	−0.1666	−0.1313	−0.0722	−0.0869	−0.1237
F_{5r}	−0.0722	−0.0820	−0.1160	−0.0722	−0.0842	−0.1068
F_{6r}	−0.0722	−0.0783	−0.0974	−0.0722	−0.0790	−0.0866
F_{7r}	−0.0722	−0.0721	−0.0758	−0.0722	−0.0714	−0.0638
F_{8r}	−0.0722	−0.0638	−0.0518	−0.0722	−0.0617	−0.0391
F_{9r}	−0.0722	−0.0535	−0.0263	−0.0722	−0.0500	−0.01316
α_1	180°	180°	180°	170°	170°	170°
α_2	160°	160°	160°	150°	150°	150°
α_3	140°	140°	140°	130°	130°	130°
α_4	120°	120°	120°	110°	110°	110°
α_5	100°	100°	100°	90°	90°	90°
α_6	80°	80°	80°	70°	70°	70°
α_7	60°	60°	60°	50°	50°	50°
α_8	40°	40°	40°	30°	30°	30°
α_9	20°	20°	20°	10°	10°	10°

Table 2.27 Harmonic analysis results of rotating magnetomotive forces of simple and STW with $q=6$

ν—Harmonic sequence number	Winding type					
	O_{16}	P_{16}	R_{16}	O_{26}	P_{26}	R_{26}
1	−0.828	−0.846	−0.868	−0.791	−0.816	−0.865
5	−0.022	−0.008	0	−0.033	−0.024	0
7	−0.020	−0.017	0	−0.017	−0.009	0
11	−0.001	0.003	0	−0.008	−0.005	0
13	−0.006	−0.006	0	−0.006	−0.003	0
17	0.002	0.004	0	−0.004	−0.002	0
19	−0.002	−0.003	0	−0.004	−0.002	0
23	0.003	0.004	0	−0.003	−0.002	0
25	0	−0.001	0	−0.003	−0.002	0
29	0.005	0.004	0	−0.004	−0.002	0
31	0.003	0.001	0	−0.005	−0.004	0
35	0.024	0.024	0.025	−0.023	−0.023	−0.025
37	−0.022	−0.023	−0.023	0.021	0.022	0.023
41	−0.003	−0.001	0	0.004	0.003	0
43	−0.003	−0.003	0	0.003	0.002	0

Table 2.28 Absolute relative values of ν-th harmonic amplitudes of rotating magnetomotive forces (f_ν) for simple and STW with $q=6$

ν–Harmonic sequence number	Winding type					
	O_{16}	P_{16}	R_{16}	O_{26}	P_{26}	R_{26}
1	1	1	1	1	1	1
5	0.027	0.009	0	0.042	0.029	0
7	0.024	0.020	0	0.021	0.011	0
11	0.001	0.004	0	0.010	0.006	0
13	0.007	0.007	0	0.008	0.004	0
17	0.002	0.005	0	0.005	0.002	0
19	0.002	0.004	0	0.005	0.002	0
23	0.004	0.005	0	0.004	0.002	0
25	0	0.001	0	0.004	0.002	0
29	0.006	0.005	0	0.005	0.002	0
31	0.004	0.001	0	0.006	0.005	0
35	0.029	0.028	0.029	0.029	0.028	0.029
37	0.027	0.027	0.026	0.027	0.027	0.027
41	0.004	0.001	0	0.005	0.004	0
43	0.004	0.004	0	0.004	0.002	0

Table 2.29 Electromagnetic efficiency factors k_{ef} of simple and STW with $q=6$

Winding type					
O_{16}	P_{16}	R_{16}	O_{26}	P_{26}	R_{26}
0.9417	0.9487	0.9563	0.9326	0.9450	0.9561

Table 2.30 Winding factors of the first and higher harmonics ($k_{w\,\nu}$) of simple and STW with $q=6$

ν—Harmonic sequence number	Winding type					
	O_{16}	P_{16}	R_{16}	O_{26}	P_{26}	R_{26}
1	0.867	0.885	0.909	0.828	0.855	0.906
5	−0.1131	−0.0423	0	−0.1708	−0.1273	0
7	−0.1448	−0.1217	0	−0.1258	−0.0679	0
11	0.00887	−0.0373	0	0.0881	0.0569	0
13	0.0753	0.0873	0	0.0796	0.0463	0
17	0.0354	0.0650	0	−0.0725	−0.0441	0
19	−0.0354	−0.0650	0	−0.0725	−0.0441	0
23	−0.0753	−0.0873	0	0.0796	0.0463	0
25	−0.00887	0.0373	0	0.0881	0.0569	0
29	0.1448	0.1217	0	−0.1258	−0.0679	0
31	0.1131	0.0423	0	0.1708	−0.1273	0
35	−0.867	−0.885	−0.909	0.828	0.855	0.906
37	0.867	0.885	0.909	−0.828	−0.855	−0.906
41	−0.1131	−0.0423	0	0.1708	0.1273	0
43	−0.1448	−0.1217	0	−0.1258	0.0679	0

2.6   Conclusions

- The fundamental (first) space harmonics of rotating magnetomotive forces in the maximum and short average pitch sinusoidal three-phase windings with optimized pulsating and rotating magnetomotive forces are practically in all cases larger than the same harmonics of the corresponding double-layer concentric windings.
- The higher space harmonics of rotating magnetomotive forces (except for slot ripples) in the maximum and short average pitch sinusoidal three-phase windings with optimized pulsating and rotating magnetomotive forces are significantly reduced compared with the same magnetomotive force harmonics of the corresponding double-layer concentric windings.
- Slot ripples of rotating magnetomotive forces in the maximum and short average pitch sinusoidal three-phase windings with optimized pulsating and rotating magnetomotive forces remain of the same magnitude as in the double-layer concentric windings.
- Higher space harmonics (except for slot ripples) of rotating magnetomotive forces in the maximum and short average pitch sinusoidal three-phase windings with optimized rotating magnetomotive forces are reduced to zero.
- The electromagnetic efficiency factors of the maximum and short average pitch sinusoidal three-phase windings with optimized pulsating and rotating magnetomotive forces are higher than the same factors of the corresponding double-layer concentric windings.
- When the number of coils in their groups is increased in the analyzed three-phase windings, the electromagnetic efficiency factors are noticeably increased.
- The calculated winding factors of the double-layer concentric and sinusoidal three-phase windings reflect the results of harmonic analysis of the stair-shaped functions of the instantaneous rotating magnetomotive forces generated in these windings.
- According to their electromagnetic properties, the short average pitch sinusoidal three-phase windings with optimized pulsating and rotating magnetomotive forces almost equals the maximum average pitch sinusoidal three-phase windings; furthermore, because of shortened coil endings the used of these windings results in savings of nonferrous metals.

Chapter 3
Filling of Sinusoidal Three-Phase Windings-Based Stator Magnetic Circuit Slots

Since numbers of coil turns in coil groups are different for all types of sinusoidal three-phase windings, for this reason, in general case, the numbers of effective conductors in the slots of magnetic circuit into which these windings are inserted are different as well. It means that the filling of these slots is unequal. The slot filling in the order of sequence periodically varies from the maximum (permitted) value down to a certain minimal value.

Coils of the largest span in coil groups of sinusoidal three-phase windings, denoted by the first number, always have the largest number of turns and, when moving toward the symmetry axes of these groups, the number of coil turns decreases. In any case, the sinusoidal three-phase windings can be implemented only as double-layer, and the coils of their single phase winding with a larger number of turns are inserted into corresponding slots of magnetic circuit along with the coils of another phase winding with a lower number of turns (Figs. 1.4 and 1.5).

3.1 Filling of Magnetic Circuit Slot of the Maximum Average Pitch STW

In the maximum average pitch sinusoidal three-phase windings, $6p$ slots of magnetic circuit are occupied by the active sides of the same phase coils (all coils of this winding with their span equal to the pole pitch), while the active sides of different phase coils are inserted into the rest $Z - 6p$ slots, where Z is the number of magnetic circuit slots, p is the number of pole pairs (Figs. 1.4, 1.9, 1.11, 1.13, 1.15). If we assume that the first slot will be filled with the active sides of the largest-span (first) coils from adjacent coil groups of any phase winding, then the second slot will be occupied by the active sides of the second coil from the same group of coils and the active sides of the q-th coil from the coil group of another phase, etc. It means that the order of insertion of active coil sides into corresponding slots according to the

© Springer International Publishing Switzerland 2016
J.J. Buksnaitis, *Sinusoidal Three-Phase Windings of Electric Machines*,
DOI 10.1007/978-3-319-42931-1_3

coil number in their group depends only on the number of coils forming particular group of coils, q. The distribution of the active coil sides into the slots of magnetic circuit according to their number in a group for the maximum average pitch sinusoidal three-phase windings in respect of the q value is presented in Table 3.1.

It can be seen from Table 3.1 that the magnetic circuit slots will have two different filling levels each when the value of q is two and three. The slots into which coils $1-1$ will be inserted will have the maximum slot filling, and the slots with coils $2-2$, $2-3$, and $3-2$ will have the minimum slot filling level. The magnetic circuit slots will have three different filling levels each when the value of q is four and five. The maximum slot filling level will be provided by $1-1$ coils of the same phase, the average filling level—by coils $2-4$ and $4-2$, $2-5$ and $5-2$, and the minimum level—by coils $3-3$, $3-4$, and $4-3$. Accordingly, the magnetic circuit slots will have four different filling levels each when the value of q is six.

The maximum average pitch sinusoidal three-phase winding slot filling levels can be determined based on the real relative values of coil turn numbers, i.e., using data from Tables 1.2–1.11, as well as data from Table 3.1. Therefore, the preliminary slot filling levels (which are expressed by their filling factors), when the pulsating magnetomotive forces of the maximum average pitch sinusoidal three-phase windings are optimized, are found using the system of Eqs. (3.1):

$$
\left\{
\begin{aligned}
\lambda_{1p1} &= 2N_{1p1}^{*}; \\
\lambda_{1p2} &= N_{1p2}^{*} + N_{1pq}^{*}; \\
\lambda_{1p3} &= N_{1p3}^{*} + N_{1p(q-1)}^{*}; \\
\hline
\end{aligned}
\right.
\tag{3.1}
$$

The preliminary slot fill factors, when the rotating magnetomotive forces of the maximum average pitch sinusoidal three-phase windings are optimized, are found using the system of Eq. (3.2):

$$
\left\{
\begin{aligned}
\lambda_{1r1} &= 2N_{1r1}^{*}; \\
\lambda_{1r2} &= N_{1r2}^{*} + N_{1rq}^{*}; \\
\lambda_{1r3} &= N_{1r3}^{*} + N_{1r(q-1)}^{*}; \\
\hline
\end{aligned}
\right.
\tag{3.2}
$$

Table 3.1 Distribution of the active coil sides into the slots of magnetic circuit according to their number in a group for the maximum average pitch STW in respect of the q value

q	Magnetic circuit slot number							
	1	2	3	4	5	6	7	...
2	$1-1$	$2-2$	$1-1$	$2-2$	$1-1$	$2-2$	$1-1$	...
3	$1-1$	$2-3$	$3-2$	$1-1$	$2-3$	$3-2$	$1-1$	...
4	$1-1$	$2-4$	$3-3$	$4-2$	$1-1$	$2-4$	$3-3$	...
5	$1-1$	$2-5$	$3-4$	$4-3$	$5-2$	$1-1$	$2-5$	...
6	$1-1$	$2-6$	$3-5$	$4-4$	$5-3$	$6-2$	$1-1$	...

Table 3.2 Preliminary and real slot fill factors for the maximum average pitch STW (P_{1q}) with the optimized pulsating magnetomotive forces

q	λ_{1p}	Magnetic circuit slot number							
		1	2	3	4	5	6	7	...
2	λ_{1pi}	0.536	0.464	0.536	0.464	0.536	0.464	0.536	...
	λ_{1pi}^{*}	1	0.866	1	0.866	1	0.866	1	...
3	λ_{1pi}	0.370	0.315	0.315	0.370	0.315	0.315	0.370	...
	λ_{1pi}^{*}	1	0.853	0.853	1	0.853	0.853	1	...
4	λ_{1pi}	0.283	0.236	0.245	0.236	0.283	0.236	0.245	...
	λ_{1pi}^{*}	1	0.8365	0.866	0.8365	1	0.8365	0.866	...
5	λ_{1pi}	0.229	0.1885	0.1971	0.1971	0.1885	0.229	0.1885	...
	λ_{1pi}^{*}	1	0.824	0.861	0.861	0.824	1	0.824	...
6	λ_{1pi}	0.1924	0.1565	0.1641	0.1666	0.1641	0.1565	0.1924	...
	λ_{1pi}^{*}	1	0.813	0.853	0.866	0.853	0.813	1	...

Table 3.3 Preliminary and real slot fill factors for the maximum average pitch STW (R_{1q}) with the optimized rotating magnetomotive forces

q	λ_{1r}	Magnetic circuit slot number							
		1	2	3	4	5	6	7	...
2	λ_{1ri}	0.464	0.536	0.464	0.536	0.464	0.536	0.464	...
	λ_{1ri}^{*}	0.866	1	0.866	1	0.866	1	0.866	...
3	λ_{1ri}	0.305	0.347	0.347	0.305	0.347	0.347	0.305	...
	λ_{1ri}^{*}	0.879	1	1	0.879	1	1	0.879	...
4	λ_{1ri}	0.228	0.254	0.263	0.254	0.228	0.254	0.263	...
	λ_{1ri}^{*}	0.866	0.965	1	0.965	0.866	0.965	1	...
5	λ_{1ri}	0.1820	0.1999	0.209	0.209	0.1999	0.1820	0.1999	...
	λ_{1ri}^{*}	0.870	0.956	1	1	0.956	0.870	0.956	...
6	λ_{1ri}	0.1516	0.1644	0.1723	0.175	0.1723	0.1644	0.1516	...
	λ_{1ri}^{*}	0.866	0.939	0.985	1	0.985	0.939	0.866	...

The real slot fill factors, λ_{1pi}^{*} and λ_{1ri}^{*}, that are associated with their permitted fill factor which equals one, are determined by dividing the preliminary fill factors of all slots λ_{1pi} or λ_{1ri} by the largest preliminary slot fill factor. Based on expressions (3.1), the list of preliminary and real slot fill factors λ_{1p} for the maximum average pitch sinusoidal three-phase windings with the optimized pulsating magnetomotive forces is created (Table 3.2).

Based on expressions (3.2), the list of preliminary and real slot fill factors λ_{1r} for the maximum average pitch sinusoidal three-phase windings with the optimized rotating magnetomotive forces is created (Table 3.3).

3.2 Fill of Magnetic Circuit Slot of the Short Average Pitch STW

In the short average pitch sinusoidal three-phase windings, all slots of magnetic circuit are occupied by the active coil sides of different phases (Figs. 1.19, 1.21, 1.23, 1.25, 1.27). If we assume that the first slot will be filled with the active sides of the first coil from phase winding U and the active sides of the q-th coil of phase winding W, the second slot—by the active sides of the second coil and $(q-1)$ coil of the same phase windings and the same coil group, etc., then the order of insertion of active coil sides into corresponding slots according to the coil number in their group depends only on the number of coils forming particular group of coils, q. The distribution of the active coil sides into the slots of magnetic circuit according to their number in a group for the short average pitch sinusoidal three-phase windings in respect of the q value is presented in Table 3.4.

It can be seen from Table 3.4 that the slots will have two different filling levels each when the value of q is three and four. The maximum slot filling level will be achieved for slots with inserted coils $1-3$ and $3-1$, $1-4$, and $4-1$, and the minimum slot filling level will be achieved for slots with inserted coils $2-2$, and $2-3$ and $3-2$. The slots will have three different filling levels each when the value of q is five and six. The maximum slot filling levels will be ensured by coils inserted into slots $1-5$ and $1-6$, the average filling levels—by coils $2-4$ and $4-2$, $2-5$, and $5-2$, and the minimum levels—by coils $3-3$, $3-4$, and $4-3$.

The short average pitch sinusoidal three-phase winding slot filling levels in their groups can be determined based on the real relative values of coil turn numbers, i.e., using data from Tables 1.12–1.21, as well as data from Table 3.4. Therefore, the

Table 3.4 Distribution of the active coil sides into the slots of magnetic circuit according to their number in a group for the short average pitch STW in respect of the q value

q	Magnetic circuit slot number							
	1	2	3	4	5	6	7	...
2	$1-2$	$2-1$	$1-2$	$2-1$	$1-2$	$2-1$	$1-2$	...
3	$1-3$	$2-2$	$3-1$	$1-3$	$2-2$	$3-1$	$1-3$	...
4	$1-4$	$2-3$	$3-2$	$4-1$	$1-4$	$2-3$	$3-2$	
5	$1-5$	$2-4$	$3-3$	$4-2$	$5-1$	$1-5$	$2-4$	...
6	$1-6$	$2-5$	$3-4$	$4-3$	$5-2$	$6-1$	$1-6$	...

Table 3.5 Preliminary λ_{2pi} and real λ_{2pi}^* slot fill factors for the short average pitch STW (P_{2q}) with the optimized pulsating magnetomotive forces

q	λ_{2p}	Magnetic circuit slot number							
		1	2	3	4	5	6	7	...
2	λ_{2pi}	0.50	0.50	0.50	0.50	0.50	0.50	0.50	...
	λ_{2pi}^*	1	1	1	1	1	1	1	...
3	λ_{2pi}	0.326	0.347	0.326	0.326	0.347	0.326	0.326	...
	λ_{2pi}^*	0.940	1	0.940	0.940	1	0.940	0.940	...
4	λ_{2pi}	0.241	0.259	0.259	0.241	0.241	0.259	0.259	...
	λ_{2pi}^*	0.932	1	1	0.932	0.932	1	1	...
5	λ_{2pi}	0.1910	0.2045	0.209	0.2045	0.1910	0.1910	0.2045	...
	λ_{2pi}^*	0.914	0.978	1	0.978	0.914	0.914	0.978	...
6	λ_{2pi}	0.1580	0.1684	0.1736	0.1736	0.1684	0.1580	0.1580	...
	λ_{2pi}^*	0.910	0.970	1	1	0.970	0.910	0.910	...

preliminary slot filling levels (which are expressed by their filling factors), when the pulsating magnetomotive forces of the short average pitch sinusoidal three-phase windings are optimized, are found using the system of Eqs. (3.3):

$$\left\{ \begin{array}{l} \lambda_{2p1} = N_{2p1}^* + N_{2pq}^*; \\ \lambda_{2p2} = N_{2p2}^* + N_{2p(q-1)}^*; \\ ----------- \end{array} \right. \tag{3.3}$$

The preliminary slot fill factors, when the rotating magnetomotive forces of the short average pitch sinusoidal three-phase windings are optimized, are found using the system of Eqs. (3.4):

$$\left\{ \begin{array}{l} \lambda_{2r1} = N_{2r1}^* + N_{2rq}^*; \\ \lambda_{2r2} = N_{2r2}^* + N_{2r(q-1)}^*; \\ ----------- \end{array} \right. \tag{3.4}$$

The real slot fill factors, λ_{2pi}^* and λ_{2ri}^*, that are associated with their permitted fill factor which equals one, are determined by dividing the preliminary fill factors of all slots λ_{2pi} or λ_{2ri} by the largest preliminary slot fill factor. Based on expressions (3.3), the list of preliminary and real slot fill factors λ_{2p} for the short average pitch sinusoidal three-phase windings with the optimized pulsating magnetomotive forces is created (Table 3.5).

Table 3.6 Preliminary λ_{2ri} and real λ_{2ri}^{*} slot fill factors for the short average pitch STW (R_{2q}) with the optimized rotating magnetomotive forces

q	λ_{2r}	Magnetic circuit slot number							
		1	2	3	4	5	6	7	...
2	λ_{2ri}	0.50	0.50	0.50	0.50	0.50	0.50	0.50	...
	λ_{2ri}^{*}	1	1	1	1	1	1	1	...
3	λ_{2ri}	0.3263	0.3474	0.3263	0.3263	0.3474	0.3263	0.3263	...
	λ_{2ri}^{*}	0.939	1	0.939	0.939	1	0.939	0.939	...
4	λ_{2ri}	0.2411	0.2588	0.2588	0.2411	0.2411	0.2588	0.2588	...
	λ_{2ri}^{*}	0.932	1	1	0.932	0.932	1	1	...
5	λ_{2ri}	0.1910	0.2045	0.209	0.2045	0.1910	0.1910	0.2045	...
	λ_{2ri}^{*}	0.914	0.978	1	0.978	0.914	0.914	0.978	...
6	λ_{2ri}	0.1580	0.1684	0.1737	0.1737	0.1684	0.1580	0.1580	...
	λ_{2ri}^{*}	0.910	0.969	1	1	0.969	0.910	0.910	...

Based on expressions (3.4), the list of preliminary and real slot fill factors λ_{2r} for the short average pitch sinusoidal three-phase windings with the optimized rotating magnetomotive forces is created (Table 3.6).

3.3 Average Filling of Magnetic Circuit Slot of the STW

The average magnetic circuit fill factors in respect of the permitted values can be calculated based on their real fill factors. Calculations of these magnitudes are performed using the following expressions:

$$\lambda_{1pav}^{*} = \left(\lambda_{1p1}^{*} + \lambda_{1p2}^{*} + \ldots + \lambda_{1pq}^{*} \right) / q; \qquad (3.5)$$

$$\lambda_{1rav}^{*} = \left(\lambda_{1r1}^{*} + \lambda_{1r2}^{*} + \ldots + \lambda_{1rq}^{*} \right) / q; \qquad (3.6)$$

$$\lambda_{2pav}^{*} = \left(\lambda_{2p1}^{*} + \lambda_{2p2}^{*} + \ldots + \lambda_{2pq}^{*} \right) / q; \qquad (3.7)$$

$$\lambda_{2rav}^{*} = \left(\lambda_{2r1}^{*} + \lambda_{2r2}^{*} + \ldots + \lambda_{2rq}^{*} \right) / q. \qquad (3.8)$$

The average magnetic circuit slot fill factors for all four types of STW with different numbers of coils in their group are listed in Table 3.7.

Table 3.7 Average magnetic circuit slot fill factors in respect of the permitted factor value, when filling STW

q	$\lambda^*_{1p\,av}$	$\lambda^*_{1r\,av}$	$\lambda^*_{2p\,av}$	$\lambda^*_{2r\,av}$
2	0.933	0.933	1	1
3	0.902	0.960	0.960	0.959
4	0.885	0.949	0.966	0.966
5	0.874	0.956	0.957	0.957
6	0.866	0.952	0.960	0.960

3.4 Conclusions

- When using sinusoidal three-phase winding of any type, the magnetic circuit slots are filled unequally, but the filling level of some slots is equal to the permitted level, and for some it is lower but close to the permitted.
- The lowest average magnetic circuit filling factors are obtained when the maximum average pitch sinusoidal three-phase windings with the optimized pulsating magnetomotive force are used.
- When using the short average pitch sinusoidal three-phase windings of any type, the same values of the average magnetic circuit slot fill factors are obtained under the same number of coils in their group.
- When the number of coils in their groups is increased, the average magnetic circuit slot fill factors noticeably decrease only when using the maximum average pitch sinusoidal three-phase windings with the optimized pulsating magnetomotive force; while these factors vary only slightly for the other types of windings.
- The reduction of the average magnetic circuit slot fill factor for the sinusoidal three-phase windings in respect of the permitted value can be associated with the reduced relative consumption of nonferrous metals in relation to the corresponding double-layer concentric (simple) three-phase windings ($(1-0.949)$ $100 = 5.1\%$).

Chapter 4
Automated Filling of STW-Based Stator Magnetic Circuit Slots

Technological schemes according to which the modern machinery that fills the windings into the stator magnetic circuit slots are operating are mostly used for such single-phase and three-phase windings which do not require lifting of their active coil sides during this technological process. This condition is met by all types of single-layer windings and double-layer maximum and short average pitch concentric (simple) windings, as well as both types of sinusoidal three-phase windings.

When filling the latter three-phase windings there is no need to lift the active coil sides, as an assumption is discarded stating that the active sides of each coil should occupy different layers in the slots of magnetic circuits, i.e., one in the upper, another—in the lower. During the insertion of these windings it is important to fulfill the condition specifying that approximately half of the active coil sides of each phase winding should be placed into the upper layers of magnetic circuit slots, while another half—into the lower layers. This means that all the active sides of any coil or even a coil group may be placed either into upper or into lower slot layers.

4.1 Technological Schemes for Mechanized Filling of Magnetic Circuit Slots of the Maximum Average Pitch STW

The technological schemes for mechanized filling of magnetic circuit slots of the maximum average pitch sinusoidal three-phase windings were created on the basis of Figs. 1.4, 1.7–1.15.

The sinusoidal three-phase winding with $q = 2$ (Figs. 1.4 and 1.7) could be inserted in a mechanized way according to the following technological scheme (Table 4.1).

© Springer International Publishing Switzerland 2016
J.J. Buksnaitis, *Sinusoidal Three-Phase Windings of Electric Machines*,
DOI 10.1007/978-3-319-42931-1_4

Table 4.1 Technological scheme for mechanized filling of magnetic circuit slots of the maximum average pitch STW with $q=2$

		Active coil side or slot number	
Technological step	Phase winding	$\mathbf{Z}$	$\mathbf{Z'}$
1	U	–	2–8 3–7
2	W	–	4–10 5–9
3	V	7–	6–12 –11
4	U	8–2 9–	–1
5	W	10–4 11–3	–
6	V	12–6 1–5	–

In the above table and other tables below the symbol $\mathbf{Z}$ denotes that the active coil sides are inserted into the upper slot layers, and $\mathbf{Z'}$—that the active coil sides are inserted into the lower slot layers.

In the first technological step, the first coil group of the phase winding U is placed into the lower layers of magnetic circuit slots 2, 3, 7, and 8. In the second step, the second group of coils of the phase winding W is placed also into the lower layers of magnetic circuit slots 4, 5, 9, and 10. In the third step, three active sides of the first coil group of the phase winding V are placed into the lower layers of slots 6, 11, 12, and one—into the upper layer of slot 7, because the lower layer of this slot is already occupied. In the fourth step, three active sides of the second coil group of the phase winding U are placed into the upper layers of slots 8, 9, 2, and one—into the lower layer of slot 1. In fifth and sixth steps, the first coil group of the phase winding W and the second coil group of the phase winding V are placed into the vacant upper layers of the corresponding slots.

The sinusoidal three-phase winding with $q=3$ (Figs. 1.8 and 1.9) could be inserted in a mechanized way according to the following technological scheme (Table 4.2).

In the first technological step, the first coil group of the phase winding U is placed into the lower layers of magnetic circuit slots 1, 2, 3, 10, 9, and 8. In the second step, the second coil group of the phase winding W is placed also into the lower layers of magnetic circuit slots 4, 5, 6, 13, 12, and 11. In the third step, four active sides of the first coil group of the phase winding V are placed into the lower layers of slots 7, 16, 15, and 14, and two—into the upper layers of slots 8 and 9, as the lower layers of these slots are already occupied. In the fourth step, four active sides of the second coil group of the phase winding U are placed into the upper layers of slots 10, 11, 12, and 1, and two—into the lower layers of slots 18 and 17. In the fifth and sixth steps, the first coil group of the phase winding W and the second coil group of the phase winding V are placed into the vacant upper layers of the corresponding slots.

Table 4.2 Technological scheme for mechanized filling of magnetic circuit slots of the maximum average pitch STW with $q=3$

Technological step	Phase winding	Active coil side or slot number	
		Z	Z'
1	U	–	1–10
			2–9
			3–8
2	W	–	4–13
			5–12
			6–11
3	V	8–	7–16
		9–	–15
			–14
4	U	10–1	–18
		11–	–17
		12–	
5	W	13–4	–
		14–3	
		15–2	
6	V	16–7	–
		17–6	
		18–5	

The sinusoidal three-phase winding with $q=4$ (Figs. 1.10 and 1.11) could be inserted in a mechanized way according to the following technological scheme (Table 4.3).

In the first technological step, the first coil group of the phase winding U is placed into the lower layers of magnetic circuit slots 1, 2, 3, 4, 13, 12, 11, and 10. In the second step, the second coil group of the phase winding W is placed also into the lower layers of magnetic circuit slots 5, 6, 7, 8, 17, 16, 15, and 14. In the third step, five active sides of the first coil group of the phase winding V are placed into the lower layers of slots 9, 21, 20, 19, and 18, and three—into the upper layers of slots 10, 11, and 12, as the lower layers of these slots are already occupied. In the fourth step, five active sides of the second coil group of the phase winding U are placed into the upper layers of slots 13, 14, 15, 16, and 1, and three—into the lower layers of slots 24, 23, and 22. In the fifth and sixth steps, the first coil group of the phase winding W and the second coil group of the phase winding V are placed into the vacant upper layers of the corresponding slots.

The sinusoidal three-phase winding with $q=5$ (Figs. 1.12 and 1.13) could be inserted in a mechanized way according to the following technological scheme (Table 4.4).

In the first technological step, the first coil group of the phase winding U is placed into the lower layers of magnetic circuit slots 1, 2, 3, 4, 5, 16, 15, 14, 13, and 12. In the second step, the second coil group of the phase winding W is placed also into the lower layers of magnetic circuit slots 6, 7, 8, 9, 10, 21, 20, 19, 18, and 17. In the third step, six active sides of the first coil group of the phase winding V are placed into the lower layers of slots 11, 26, 25, 24, 23, and 22, and four—into

Table 4.3 Technological scheme for mechanized filling of magnetic circuit slots of the maximum average pitch STW with $q=4$

Technological step	Phase winding	Active coil side or slot number	
		Z	Z´
1	U	–	1–13 2–12 3–11 4–10
2	W	–	5–17 6–16 7–15 8–14
3	V	10– 11– 12–	9–21 –20 –19 –18
4	U	13–1 14– 15– 16–	–24 –23 –22
5	W	17–5 18–4 19–3 20–2	–
6	V	21–9 22–8 23–7 24–6	–

the upper layers of slots 12, 13, 14, and 15, as the lower layers of these slots are already occupied. In the fourth step, six active sides of the second coil group of the phase winding U are placed into the upper layers of slots 16, 17, 18, 19, 20, and 1, and four—into the lower layers of slots 30, 29, 28, and 27. In the fifth and sixth steps, the first coil group of the phase winding W and the second coil group of the phase winding V are placed into the vacant upper layers of the corresponding slots.

The sinusoidal three-phase winding with $q=6$ (Figs. 1.14 and 1.15) could be inserted in a mechanized way according to the following technological scheme (Table 4.5).

In the first technological step, the first coil group of the phase winding U is placed into the lower layers of magnetic circuit slots 1, 2, 3, 4, 5, 6, 19, 18, 17, 16, 15, and 14. In the second step, the second coil group of the phase winding W is placed also into the lower layers of magnetic circuit slots 7, 8, 9, 10, 11, 12, 25, 24, 23, 22, 21, and 20. In the third step, seven active sides of the first coil group of the

Table 4.4 Technological scheme for mechanized filling of magnetic circuit slots of the maximum average pitch STW with $q = 5$

Technological step	Phase winding	Active coil side or slot number	
		Z	Z'
1	U	–	1–16
			2–15
			3–14
			4–13
			5–12
2	W	–	6–21
			7–20
			8–19
			9–18
			10–17
3	V	12–	11–26
		13–	–25
		14–	–24
		15–	–23
			–22
4	U	16–1	–30
		17–	–29
		18–	–28
		19–	–27
		20	
5	W	21–6	–
		22–5	
		23–4	
		24–3	
		25–2	
6	V	26–11	–
		27–10	
		28–9	
		29–8	
		30–7	

phase winding V are placed into the lower layers of slots 13, 31, 30, 29, 28, 27, and 26, and five—into the upper layers of slots 14, 15, 16, 17, and 18, as the lower layers of these slots are already occupied. In the fourth step, seven active sides of the second coil group of the phase winding U are placed into the upper layers of slots 19, 20, 21, 22, 23, 24, and 1, and five—into the lower layers of slots 36, 35, 34, 33, and 32. In the fifth and sixth steps, the first coil group of the phase winding W and the second coil group of the phase winding V are placed into the vacant upper layers of the corresponding slots.

Table 4.5 Technological scheme for mechanized filling of magnetic circuit slots of the maximum average pitch STW with $q=6$

Technological step	Phase winding	Active coil side or slot number	
		Z	**Z'**
1	U	–	1–19
			2–18
			3–17
			4–16
			5–15
			6–14
2	W	–	7–25
			8–24
			9–23
			10–22
			11–21
			12–20
3	V	14–	13–31
		15–	–30
		16–	–29
		17–	–28
		18–	–27
			–26
4	U	19–1	–36
		20–	–35
		21–	–34
		22–	–33
		23–	–32
		24–	
5	W	25–7	–
		26–6	
		27–5	
		28–4	
		29–3	
		30–2	
6	V	31–13	–
		32–12	
		33–11	
		34–10	
		35–9	
		36–8	

4.2 Technological Schemes for Mechanized Filling of Magnetic Circuit Slots of the Short Average Pitch STW

The technological schemes for mechanized filling of magnetic circuit slots of the short average pitch sinusoidal three-phase windings were created on the basis of Figs. 1.18–1.27.

Table 4.6 Technological scheme for mechanized filling of magnetic circuit slots of the short average pitch STW with $q=2$

Technological step	Phase winding	Active coil side or slot number	
		Z	Z'
1	U	–	1–6 2–5
2	W	–	3–8 4–7
3	V	5– 6–	–10 –9
4	U	7– 8–	–12 –11
5	W	9–2 10–1	–
6	V	11–4 12–3	–

The sinusoidal three-phase winding with $q=2$ (Figs. 1.18 and 1.19) could be inserted in a mechanized way according to the following technological scheme (Table 4.6).

In the first technological step, the first coil group of the phase winding U is placed into the lower layers of magnetic circuit slots 1, 2, 6, and 5. In the second step, the second coil group of the phase winding W is placed also into the lower layers of magnetic circuit slots 3, 4, 8, and 7. In the third step, two active sides of the first coil group of the phase winding V are placed into the upper layers of slots 5 and 6, as the lower layers of these slots are already occupied, and two active sides are placed into the lower layers of slots 10 and 9. In the fourth step, two active sides of the second coil group of the phase winding U are placed into the upper layers of slots 7 and 8, and two—into the lower layers of slots 12 and 11. In the fifth and sixth steps, the first coil group of the phase winding W and the second coil group of the phase winding V are placed into the vacant upper layers of the corresponding slots.

The sinusoidal three-phase winding with $q=3$ (Figs. 1.20 and 1.21) could be inserted in a mechanized way according to the following technological scheme (Table 4.7).

In the first technological step, the first coil group of the phase winding U is placed into the lower layers of magnetic circuit slots 1, 2, 3, 9, 8, and 7. In the second step, the second coil group of the phase winding W is placed also into the lower layers of magnetic circuit slots 4, 5, 6, 12, 11, and 10. In the third step, three active sides of the first coil group of the phase winding V are placed into the upper layers of slots 7, 8, and 9, as the lower layers of these slots are already occupied, and three—into the lower layers of slots 15, 14, and 13. In the fourth step, three active sides of the second coil group of the phase winding U are placed into the upper layers of slots 10, 11, and 12, and three—into the lower layers of slots 18, 17, and 16. In the fifth and sixth steps, the first coil group of the phase winding W and the second coil group of the phase winding V are placed into the vacant upper layers of the corresponding slots.

Table 4.7 Technological scheme for mechanized filling of magnetic circuit slots of the short average pitch STW with $q = 3$

Technological step	Phase winding	Active coil side or slot number	
		Z	Z′
1	U	–	1–9 2–8 3–7
2	W	–	4–12 5–11 6–10
3	V	7– 8– 9–	–15 –14 –13
4	U	10– 11– 12–	–18 –17 –16
5	W	13–3 14–2 15–1	–
6	V	16–6 17–5 18–4	–

Table 4.8 Technological scheme for mechanized filling of magnetic circuit slots of the short average pitch STW with $q = 4$

Technological step	Phase winding	Active coil side or slot number	
		Z	Z′
1	U	–	1–12 2–11 3–10 4–9
2	W	–	5–16 6–15 7–14 8–13
3	V	9– 10– 11– 12–	–20 –19 –18 –17
4	U	13– 14– 15– 16–	–24 –23 –22 –21
5	W	17–4 18–3 19–2 20–1	–
6	V	21–8 22–7 23–6 24–5	–

Table 4.9 Technological scheme for mechanized filling of magnetic circuit slots of the short average pitch STW with $q=5$

Technological step	Phase winding	Active coil side or slot number	
		Z	Z'
1	U	–	1–15 2–14 3–13 4–12 5–11
2	W	–	6–20 7–19 8–18 9–17 10–16
3	V	11– 12– 13– 14– 15–	–25 –24 –23 –22 –21
4	U	16– 17– 18– 19– 20–	–30 –29 –28 –27 –26
5	W	21–5 22–4 23–3 24–2 25–1	–
6	V	26–10 27–9 28–8 29–7 30–6	–

The sinusoidal three-phase winding with $q=4$ (Figs. 1.22 and 1.23) could be inserted in a mechanized way according to the following technological scheme (Table 4.8).

In the first technological step, the first coil group of the phase winding U is placed into the lower layers of magnetic circuit slots 1, 2, 3, 4, 12, 11, 10, and 9. In the second step, the second coil group of the phase winding W is placed also into the lower layers of magnetic circuit slots 5, 6, 7, 8, 16, 15, 14, and 13. In the third step, four active sides of the first coil group of the phase winding V are placed into the upper layers of slots 9, 10, 11, and 12, as the lower layers of these slots are already occupied, and four—into the lower layers of slots 20, 19, 18, and 17. In the fourth step, four active sides of the second coil group of the phase winding U are placed into the upper layers of slots 13, 14, 15, and 16, and four—into the lower layers of slots 24, 23, 22, and 21. In the fifth and sixth steps, the first coil group of

Table 4.10 Technological scheme for mechanized filling of magnetic circuit slots of the short average pitch STW with $q=6$

Technological step	Phase winding	Active coil side or slot number	
		Z	Z'
1	U	–	1–18
			2–17
			3–16
			4–15
			5–14
			6–13
2	W	–	7–24
			8–23
			9–22
			10–21
			11–20
			12–19
3	V	13–	–30
		14–	–29
		15–	–28
		16–	–27
		17–	–26
		18–	–25
4	U	19–	–36
		20–	–35
		21–	–34
		22–	–33
		23–	–32
		24–	–31
5	W	25–6	–
		26–5	
		27–4	
		28–3	
		29–2	
		30–1	
6	V	31–12	–
		32–11	
		33–10	
		34–9	
		35–8	
		36–7	

the phase winding W and the second coil group of the phase winding V are placed into the vacant upper layers of the corresponding slots.

The sinusoidal three-phase winding with $q=5$ (Figs. 1.24 and 1.25) could be inserted in a mechanized way according to the following technological scheme (Table 4.9).

In the first technological step, the first coil group of the phase winding U is placed into the lower layers of magnetic circuit slots 1, 2, 3, 4, 5 15, 14, 13, 12, and 11. In the second step, the second coil group of the phase winding W is placed also into the lower layers of magnetic circuit slots 6, 7, 8, 9, 10, 20, 19, 18, 17, and 16.

In the third step, five active sides of the first coil group of the phase winding V are placed into the upper layers of slots 11, 12, 13, 14, and 15, as the lower layers of these slots are already occupied, and five—into the lower layers of slots 25, 24, 23, 22, and 21. In the fourth step, five active sides of the second coil group of the phase winding U are placed into the upper layers of slots 16, 17, 18, 19, and 20, and five—into the lower layers of slots 30, 29, 28, 27, and 26. In the fifth and sixth steps, the first coil group of the phase winding W and the second coil group of the phase winding V are placed into the vacant upper layers of the corresponding slots.

The sinusoidal three-phase winding with $q=6$ (Figs. 1.26 and 1.27) could be inserted in a mechanized way according to the following technological scheme (Table 4.10).

In the first technological step, the first coil group of the phase winding U is placed into the lower layers of magnetic circuit slots 1, 2, 3, 4, 5, 6, 18, 17, 16, 15, 14, and 13. In the second step, the second coil group of the phase winding W is placed also into the lower layers of magnetic circuit slots 7, 8, 9, 10, 11, 12, 24, 23, 22, 21, 20, and 19. In the third step, six active sides of the first coil group of the phase winding V are placed into the upper layers of slots 13, 14, 15, 16, 17, and 18, as the lower layers of these slots are already occupied, and six—into the lower layers of slots 30, 29, 28, 27, 26, and 25. In the fourth step, six active sides of the second coil group of the phase winding U are placed into the upper layers of slots 19, 20, 21, 22, 23, and 24, and six—into the lower layers of slots 36, 35, 34, 33, 32, and 31. In the fifth and sixth steps, the first coil group of the phase winding W and the second coil group of the phase winding V are placed into the vacant upper layers of the corresponding slots.

4.3 Conclusions

- The maximum and short average pitch sinusoidal three-phase windings could be inserted into the slots of magnetic circuit in a mechanized way according to the presented technological schemes, without lifting active coil sides.
- In the presented technological schemes for the mechanized filling of the magnetic circuit slots of the maximum and short average pitch sinusoidal three-phase windings, one coil group of the corresponding phase winding is inserted at a time in each step of these schemes, where the coils forming a particular group do not have to be additionally connected in series after their insertion.
- In the presented technological schemes for the mechanized filling of the magnetic circuit slots of the maximum and short average pitch sinusoidal three-phase windings, only the second groups of coils from the phase winding W occupy the bottom slot layers and the first groups occupy the top layers. The first coil groups of the phase winding U occupy the bottom slot layers, while $(q+1)$ active coil sides from the second coil groups are placed into the upper layers of slots and $(q-1)$ active coil sides are placed into the lower layers. For the phase winding V, $(q+1)$ active coil sides from the first coil groups are placed into the lower layers of slots and $(q-1)$ active coil sides are placed into the upper layers, while the second coil groups occupy the upper layers of slots.

- In the presented technological schemes for the mechanized filling of the magnetic circuit slots of the short average pitch sinusoidal three-phase windings, only the second coil groups of the phase winding W occupy the lower layers and the first coil groups occupy the upper layers of slots. The first coil groups of the phase winding U occupy the lower layers of slots, while q active coil sides from the second coil groups are placed into the upper layers of slots and q active coil sides are placed into the lower layers. For the phase winding V, q active coil sides from the first coil groups are placed into the lower layers of slots and q active coil sides are placed into the upper layers, while the second coil groups occupy the upper layers of slots.

Chapter 5
Power Parameters of Induction Motors and Electromagnetic Efficiency of Their Windings

A large part of the consumed electrical energy is transformed into mechanical energy in the electric drives of various machines and devices. Three-phase cage rotor induction motors are used to transform the energy in these drives because their construction is simple, they are the most reliable during exploitation, have the least relative weight and are the least expensive. The three-phase stator winding is one of the most important construction parts of these motors. The main energy interchange processes take place jointly in this winding and in the magnetic circuit, and therefore they essentially determine the overall operation of the motor.

When the electric currents forming the symmetric three-phase current system flow through the three-phase winding of induction motor, they create non-sinusoidal magnetic fields which move in space and periodically change their shape in the course of time. Usually only odd space harmonics except for the multiples of three exist in the harmonic spectrum of these non-sinusoidal magnetic fields.

There are many different constructions of the three-phase windings of induction motors and each of them has distinctive parameters. Therefore the harmonic spectrum of the magnetic fields created by these windings and at the same time the electromagnetic properties differ, and thus they determine the power parameters and operation quality of induction motors. The electromagnetic efficiency factor is used to evaluate electromagnetic properties of three-phase windings.

The aim of this chapter is to perform a theoretical analysis of electromagnetic efficiencies of two types of three-phase windings and to relate them theoretically and experimentally to the power parameters of particular induction motors.

5.1 Object of Research

The 1.5 kW three-phase induction motor of standard size with the preformed single-layer winding and the same motor with the stator winding replaced with the sinusoidal winding is investigated here. Common stator parameters for both motors are the

© Springer International Publishing Switzerland 2016   87
J.J. Buksnaitis, *Sinusoidal Three-Phase Windings of Electric Machines*,
DOI 10.1007/978-3-319-42931-1_5

following: number of phases $m=3$; number of stator magnetic circuit slots $Z=24$; number of poles $2p=2$; number of stator slots (coils) per pole per phase $q=Z/(2p\,m)=24/(2\times3)=4$; pole pitch $\tau=Z/2p==24/2=12$; slot span expressed in electrical degrees $\alpha=360°p/Z=360°\times1/24=15°$. The relative value of number of turns of any coil for the preformed single-layer winding sections with four coils is $N_1^*=1/q=1/4=0.25$. Relative values of number of turns of any section in sinusoidal winding calculated according to corresponding formulas are obtained (Table 1.9): $N_{1r1}^*=0.1140$; $N_{1r2}^*=0.1862$; $N_{1r3}^*=0.1317$; $N_{1r4}^*=0.0681$.

Distribution of elements of the analyzed windings is given in Tables 5.1 and 5.2.

Electrical circuit layout of the preformed single-layer winding is created according to the data presented in Table 5.1 (Fig. 5.1a).

Electric circuit layout of the sinusoidal three-phase winding is formed according to the data presented in Table 5.2 (Fig. 5.2a).

The relative magnitudes of the instantaneous values of electric currents in both windings at the time $t=0$ are $i_U^*=\sin0°=0$; $i_V^*=\sin120°=0.866$; $i_W^*=\sin240°=-0.866$. The conditional magnetomotive force changes " $F=i^*N^*$ in the slots of magnetic circuit of the stator at time $t=0$ (Tables 5.3 and 5.4) are calculated according to the determined number of coil turns and relative magnitudes of electric currents by using the layouts of electric circuits of the analyzed windings.

The space distributions of rotating magnetomotive force at the defined moment of time are determined according to the results from Tables 5.3 and 5.4 (Figs. 5.1b and 5.2b).

Table 5.1 Distribution of elements of preformed single-layer three-phase winding

Phase alteration	U1	W2	V1	U2	W1	V2
Number of coils in a section	4	4	4	4	4	4
Slot no.	1; 2; 3; 4	5; 6; 7; 8	9; 10; 11; 12	13; 14; 15; 16	17; 18; 19; 20	21; 22; 23; 24

Table 5.2 Distribution of elements of sinusoidal three-phase winding

Phase alteration		U1	W2	V1	U2	W1	V2
Number of coils in a section		4	4	4	4	4	4
Slot no.	Z	1; 2; 3; 4	5; 6; 7; 8	9; 10; 11; 12	13; 14; 15; 16	17; 18; 19; 20	21; 22; 23; 24
	Z	10; 11; 12; 13	14; 15; 16; 17	18; 19; 20; 21	22; 23; 24; 1	2; 3; 4; 5	6; 7; 8; 9

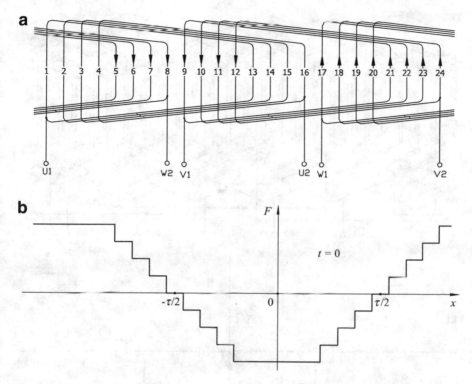

Fig. 5.1 Electrical circuit layout of preformed single-layer three-phase winding with $q=4$ (**a**), and the distribution of its rotating magnetomotive force at $t=0$ (**b**)

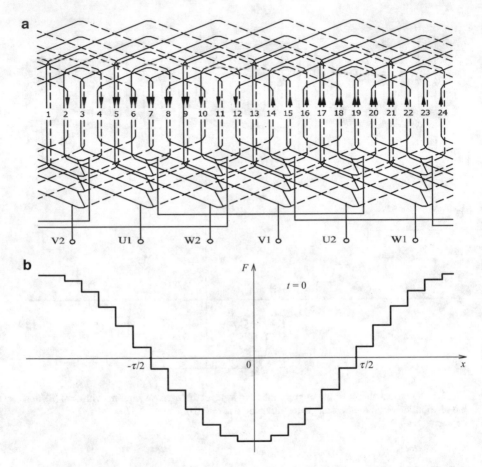

Fig. 5.2 Electrical circuit layout of sinusoidal three-phase winding with $q=4$ (**a**), and the distribution of its rotating magnetomotive force at $t=0$ (**b**)

5.2   Research Results

According to the expression (2.3) and determined parameters of rotating magnetomotive force half-period ($k=4$; $F_{1r} = -0.2165$; $F_{2r} = -0.2165$; $F_{3r} = -0.2165$; $F_{4r} = -0.2165$; $\alpha_1 = 165°$; $\alpha_2 = 135°$; $\alpha_3 = 105°$; $\alpha_4 = 75°$) the harmonic analysis of instantaneous rotating magnetomotive force function (Fig. 5.1b) of preformed single-layer three-phase winding (Fig. 5.1a) was completed and relative magnitudes of its space harmonics were calculated (Table 5.5).

According to expression (2.3) and previously determined parameters of rotating magnetomotive force half-period ($k=6$; $F_{1r} = -0.1140$; $F_{2r} = -0.2203$; $F_{3r} = -0.1975$; $F_{4r} = -0.1613$; $F_{5r} = -0.1140$; $F_{6r} = -0.0590$; $\alpha_1 = 180°$; $\alpha_2 = 150°$; $\alpha_3 = 120°$; $\alpha_4 = 90°$; $\alpha_5 = 60°$; $\alpha_6 = 30°$) the harmonic analysis of the instantaneous rotating magnetomotive force function (Fig. 5.2b) of the sinusoidal

Table 5.3 Conditional changes of magnetomotive force in slots of preformed single layer three-phase winding at $t=0$

Slot no.	1	2	3	4	5	6	7	8	9	10	11
ΔF	0	0	0	0	-0.216	-0.216	-0.216	-0.216	-0.216	-0.216	-0.216

12	13	14	15	16	17	18	19	20	21	22	23	24
-0.216	0	0	0	0	0.216	0.216	0.216	0.216	0.216	0.216	0.216	0.216

Table 5.4 Conditional changes of magnetomotive force in slots of sinusoidal three-phase winding at $t=0$

Slot no.	1	2	3	4	5	6	7	8	9	10
ΔF	0	-0.0590	-0.1140	-0.1613	-0.1975	-0.220	-0.228	-0.220	-0.1975	-0.1613

11	12	13	14	15	16	17	18	19	20	21	22
-0.114	-0.059	0	0.059	0.114	0.1613	0.1975	0.22	0.228	0.22	0.1975	0.1613

Table 5.5 Results of harmonic analysis of the instantaneous rotating magnetomotive force function of the preformed single-layer three-phase winding with $q=4$ and relative magnitudes of its space harmonics

v	1	5	7	11	13	17	19	23	25
F_{mv}	-0.914	0.0390	0.0210	-0.0110	-0.009	0.0090	0.0100	-0.0400	0.0370
f_v	1	0.0429	0.0235	0.01197	0.0101	0.00968	0.0113	0.0435	0.0400

Table 5.6 Results of harmonic analysis of the instantaneous rotating magnetomotive force function of the sinusoidal three-phase winding with $q=4$ and relative magnitudes of its space harmonics

v	1	5	7	11	13	17	19	23	25
F_{mv}	-0.871	0	0	0	0	0	0	0.0380	-0.035
f_v	1	0	0	0	0	0	0	0.0436	0.0402

three-phase winding (Fig. 5.2a) was performed and relative magnitudes of its space harmonics were calculated (Table 5.6).

According to expression (2.4) the respective electromagnetic efficiency factors k_{ef} of the preformed single layer and sinusoidal three-phase windings with $q=4$ are calculated: $k_{ef1}=0.9139$; $k_{ef2}==0.9335$. The obtained electromagnetic efficiency factor of the sinusoidal three-phase winding is 2.14 % higher than in case of preformed single-layer winding.

Experimental tests of the standard size induction motor with the researched preformed single-layer winding and motor with stator winding replaced with sinusoidal three-phase winding (under no-load and load conditions) were performed and power parameters of analyzed motors were calculated according to received results using the segregated-losses method (Tables 5.7 and 5.8).

In Tables 5.7 and 5.8, I_1—phase current of stator winding; P_1—consumed power; n—rotational speed of rotor; T—electromagnetic torque; ΣP—total power losses of

Table 5.7 Experimental and calculation results of the standard size induction motor with single-layer preformed winding

No.	I_1, A	P_1, W	n, min^{-1}	T, Nm	ΣP W	P_2, W	η	$\cos \varphi$
1	1.75	405	2983	0.586	315	90	0.222	0.351
2	2.03	840	2961	1.93	333	507	0.604	0.627
3	2.30	1110	2945	2.71	361	749	0.675	0.731
4	2.70	1410	2924	3.61	402	1008	0.715	0.791
5	3.13	1725	2899	4.50	457	1268	0.735	0.835
6	3.65	2100	2870	5.55	535	1565	0.745	0.872
7	4.13	2370	2851	6.23	610	1760	0.743	0.869
8	4.98	2805	2810	7.38	741	2064	0.736	0.853

Table 5.8 Experimental and calculation results of the induction motor with stator winding replaced with the sinusoidal three-phase winding

No.	I_1, A	P_1, W	n, min^{-1}	T, Nm	ΣP, W	P_2, W	η	$\cos \varphi$
1	1.80	385	2948	0.83	232	153	0.397	0.324
2	2.20	1215	2923	3.31	298	917	0.755	0.837
3	2.50	1440	2902	3.95	337	1103	0.766	0.873
4	3.03	1800	2868	4.92	419	1381	0.767	0.900
5	3.35	1995	2847	5.45	472	1523	0.763	0.902
6	3.60	2145	2825	5.83	522	1623	0.757	0.903
7	3.95	2345	2798	6.33	596	1749	0.746	0.899
8	4.50	2655	2756	7.08	718	1937	0.730	0.894

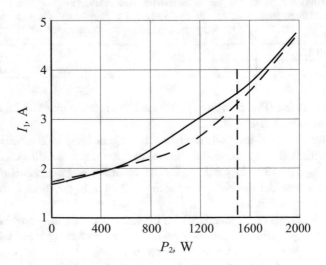

Fig. 5.3 Diagrams of function $I_1 = f(P_2)$ of the standard size motor (——) and motor with the stator winding replaced (– – –)

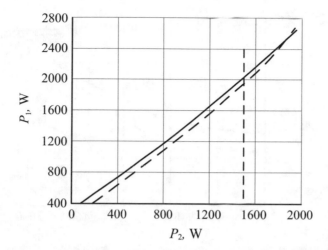

Fig. 5.4 Diagrams of function $P_1=f(P_2)$ of the standard size motor (———) and motor with the stator winding replaced (– – –)

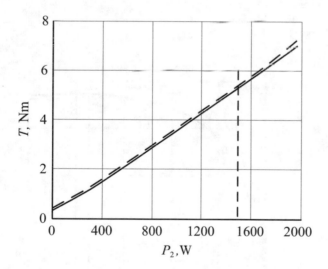

Fig. 5.5 Diagrams of function $T=f(P_2)$ of the standard size motor (———) and motor with the stator winding replaced (– – –)

motor; P_2—useful power; η—efficiency; $\cos\varphi$—power factor (Figs. 5.3, 5.4, 5.5, 5.6, 5.7, and 5.8).

After comparing the experimental and calculation results under indicated load from Tables 1.7 and 1.8, it is apparent that in case of induction motor with stator winding replaced with the sinusoidal three-phase winding the phase current of the stator winding decreased by 6.9%, power taken from electric grid decreased by 5.0%, power losses decreased by 11.7%, efficiency factor increased by 2.4% and power factor increased by 3.4%.

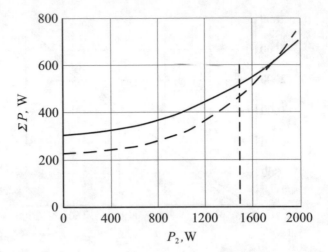

Fig. 5.6 Diagrams of function $\Sigma P = f(P_2)$ of the standard size motor (———) and motor with the stator winding replaced (– – –)

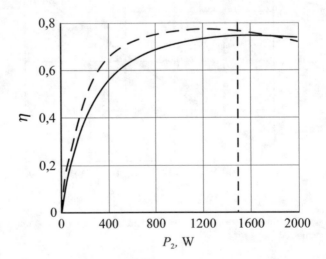

Fig. 5.7 Diagrams of function $\eta = f(P_2)$ of the standard size motor (———) and motor with the stator winding replaced (– – –)

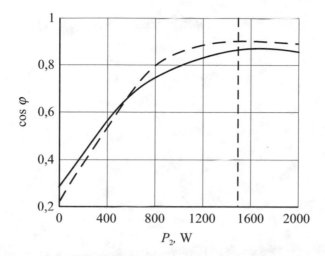

Fig. 5.8 Diagrams of function $\cos \varphi = f(P_2)$ of the standard size motor (———) and motor with the stator winding replaced (– – –)

5.3 Conclusions

- Electromagnetic properties of the three-phase windings can be evaluated by performing harmonic analysis of the rotating magnetomotive force created by these windings and by calculating electromagnetic efficiency factors based on the results of this analysis.
- It was determined theoretically that the electromagnetic efficiency factor of the preformed single-layer three-phase winding $k_{ef1} = 0.9139$ and of sinusoidal three-phase winding — $k_{ef2} = 0.9335$, i.e., 2.14 % higher than the same factor of the first winding.
- In case of induction motor with the sinusoidal three-phase winding under the indicated load, the phase current of the stator winding decreased by 6.9 %, power taken from electric grid decreased by 5.0 %, power losses decreased by 11.7 %, efficiency factor increased by 2.4 % and power factor increased by 3.4 % compared to the respective power parameters of the same motor with the preformed single-layer winding calculated under the same load.
- Induction motors having the value of the stator winding electromagnetic efficiency factor closer to 1 have better power-related parameters.

Bibliography

1. Fitzgerald, A.E., Kingsley, C., Kusko, A.: Electric machinery. McGraw-Hill Book Comp, New York (1971)
2. Slemon, G.R., Straughen, A.: Electric machines. Addison-Wesley Publ. Comp, Reading, MA (1980)
3. Krause, P.C., Wasynczuk, O., Sudhoff, S.D.: Analysis of electric machinery. The Institute of Electrical and Electronics Engineers, McGraw-Hill, New York (1995). 564 p
4. Chapman, S.J.: Electric machinery and power system fundamentals. McGraw-Hill, New York (2001). 333 p
5. Thomas, J.B.: Electromechanics of particles. Cambridge University Press, Cambridge, UK (2005). 265 p
6. Saurabh, K.M., Ahmad, S.K., Yatendra, P.S.: Electromagnetics for electrical machines. CRC Press/Taylor & Francis Group, Boca Raton, FL (2015). 421 p
7. Ivanov-Smolenskyi, A.: Electrical machines, vol. 1, 2. MIR Publishers, Moscow (1988). 400–464 p. (in Russian)
8. Livsic-Garik, M.: Windings of alternating current electrical machines, 766 p. (in Russian), Translated from English, Moscow Power Engineering Institute (MPEI) (1959)
9. Kučera, J., Gapl, I.: Windings of rotating electrical machines. Translated from Czech, 982 p., (in Russian), Czech Academy of Sciences, Prague, (1963)
10. Zerve, G.K.: Windings of electrical machines. Energoatomizdat Publishers, Leningrad (1989). 399 p. (in Russian)
11. Lopuchina, E.M., Somichina, G.S.: Calculations of single-phase and three-phase current low power induction motors. Gosenergoizdat Publishers, Moscow (1961). 245 p. (in Russian)
12. Smilgevičius, A.: Harmonic composition of magnetomotive force of concentric distributed windings. Electron. Electr. Eng. 2(44), 26–29 (2003). Technology, Kaunas, (in Lithuanian)
13. Buksnaitis, J.: The investigation of two-layer three-phase winding applied to mechanized laying. Electron. Electr. Eng. 6(48), 52–56 (2003). Technology, Kaunas (in Lithuanian)
14. Buksnaitis, J.: Substantiation and research of sinusoidal three-phase winding. Power Engineering Vilnius: Publishing House of the Lithuanian Academy of Sciences, 2, 20–27 (2004) (in Lithuanian)
15. Buksnaitis, J.: Research of formation sinusoidal three-phase winding. Electron. Electr. Eng. 1(50), 46–51 (2004). Technology, Kaunas (in Lithuanian)
16. Buksnaitis, J.: Sinusoidal three-phase winding with maximal average span. Electron. Electr. Eng. 4(60), 45–49 (2005). Technology, Kaunas (in Lithuanian)
17. Buksnaitis, J.: The research of the sinusoidal three-phase windings. Electron. Electr. Eng. 6(70), 23–28 (2006). Technology, Kaunas

© Springer International Publishing Switzerland 2016

J.J. Buksnaitis, *Sinusoidal Three-Phase Windings of Electric Machines*,

DOI 10.1007/978-3-319-42931-1

18. Buksnaitis, J.: New approach for evaluation of electromagnetic properties of three-phase windings. Electron. Electr. Eng. **3**(75), 31–36 (2007). Technology, Kaunas
19. Buksnaitis, J.: Research of electromagnetic parameters of sinusoidal three-phase windings. Electron. Electr. Eng. **8**(80), 77–82 (2007). Technology, Kaunas
20. Buksnaitis, J.: Power indexes of induction motors and electromagnetic efficiency their windings. Electron. Electr. Eng. **4**(100), 11–14 (2010). Technology, Kaunas

Index

© Springer International Publishing Switzerland 2016
J.J. Buksnaitis, *Sinusoidal Three-Phase Windings of Electric Machines*,
DOI 10.1007/978-3-319-42931-1

Printed in the United States
By Bookmasters